Lie Symmetry Analysis of Fractional Differential Equations

Lie Symmetry Analysis of Fractional Differential Equations

Mir Sajjad Hashemi
University of Bonab
Dumitru Baleanu
Institute of Space Sciences, Romania
and
Cankaya University, Turkey

CRC Press is an imprint of the
Taylor & Francis Group, an informa business
A CHAPMAN & HALL BOOK

First edition published 2020
by CRC Press
6000 Broken Sound Parkway NW, Suite 300, Boca Raton, FL 33487-2742

and by CRC Press
2 Park Square, Milton Park, Abingdon, Oxon, OX14 4RN

ISBN: 978-0-367-44186-9 (hbk)
ISBN: 978-1-003-00855-2 (ebk)

Typeset in CMR
by Nova Techset Private Limited, Bengaluru & Chennai, India

Visit the Taylor & Francis Web site at
http://www.taylorandfrancis.com

and the CRC Press Web site at
http://www.crcpress.com

*To **Elham Darvishi** and **Mihaela Cristina** for their patience and support.*

Contents

Preface

The Lie method (the terminology "the Lie symmetry analysis" and "the group analysis" are also used) is based on finding Lie's symmetries of a given differential equation and using the symmetries obtained for the construction of exact solutions. The method was created by the prominent Norwegian mathematician Sophus Lie in the 1880s. It should be pointed out that Lie's works on application Lie groups for solving PDEs were almost forgotten during the first half of the 20th century. In the end of the 1950s, L.V. Ovsiannikov, inspired by Birkhoff's works devoted to application of Lie groups in hydrodynamics, rewrote Lie's theory using modern mathematical language and published a monograph in 1962, which was the first book (after Lie's works) devoted fully to this subject. The Lie method was essentially developed by L.V. Ovsiannikov, W.F. Ames, G. Bluman, W.I. Fushchych, N. Ibragimov, P. Olver, and other researchers in the 1960s–1980s. Several excellent textbooks devoted to the Lie method were published during the last 30 years; therefore one may claim that it is the well-established theory at the present time. Notwithstanding the method still attracts the attention of many researchers and new results are published on a regular basis. In particular, solving the so-called problem of group classification (Lie symmetry classification) still remains a highly nontrivial task and such problems are not solved for several classes of PDEs arising in real world applications.

Fractional calculus is an emerging field with ramifications and excellent applications in several fields of science and engineering. During the first attempt to think about what is derivative of order 1/2, stated by Leibniz in 1695, it was considered as a paradox as mentioned by L'Hopital. Since then the trajectory of the fractional calculus passed by several periods of intensive development both in pure and applied sciences. During the last few decades the fractional calculus has been associated with the power law effects and its various applications. It is a natural question to ask if the fractional calculus, as a non-local one, can produce new results within the well-established field of Lie symmetries and their applications. In fact the fractional calculus was associated with the dissipative phenomena; therefore it is a delicate question: can we have conservation laws for fractional differential equations associated to real world models?

In our book we try to answer to this vital question by analyzing, mainly, some different aspects of fractional Lie symmetries and related conservation laws. Also, finding the exact solutions of a given fractional partial differential equation is not an easy task but we present this issue in our book. The book includes also a generalization of Lie symmetries for fractional integro-differential equations. Nonclassical Lie symmetries are discussed for fractional differential equations. Moreover, the invariant subspace method is considered to find the exact solutions of some fractional differential equations. In the present book, we assume the reader to be familiar with preliminaries of Lie symmetries for integer order differential equations.

The structure of the book is as follows. The book consists of five chapters as it is given below. In order to make the readers understand easily the topic of Lie symmetries and their applications, in Chapter 1, we show briefly the classical, nonclassical symmetries and the conservation laws of some specific problems with integer order. Next, in Chapter 2, we discuss the Lie symmetries of fractional differential equations and exact solutions with invariant subspace methods. Chapter 3 focuses on Lie symmetries of fractional integro-differential equations. The nonclassical Lie symmetry analysis of fractional differential equations is described in Chapter 4. The self-adjointness and conservation laws of fractional differential equations are considered in Chapter 5.

We believe that our book will be useful for PhD and postdoc graduates as well as for all mathematicians and applied researchers who use the powerful concept of Lie symmetries.

Authors

Mir Sajjad Hashemi is associate professor at the University of Bonab, Iran. His fields of interest include fractional differential equations, Lie symmetry method, geometric integration, approximate and analytical solutions of differential equations and soliton theory.

Dumitru Baleanu is professor at the Institute of Space Sciences, Magurele-Bucharest, Romania and a visiting staff member at the Department of Mathematics, Cankaya University, Ankara, Turkey. His fields of interest include fractional dynamics and its applications in science and engineering, fractional differential equations, discrete mathematics, mathematical physics, soliton theory, Lie symmetry, dynamic systems on time scales and the wavelet method and its applications.

Chapter 1

Lie symmetry analysis of integer order differential equations

This chapter deals with the classical and nonclassical Lie symmetry analysis of some integer order differential equations. Finding the exact solutions of differential equations is an interesting field of many researchers. The Lie symmetry method is one of the most powerful and popular ones which can analyze different types of differential equations. In the last decade, various interesting textbooks have discussed the Lie symmetry analysis of integer order differential and integro-differential equations, e.g., [59, 94, 131, 150, 29]. Various classical concerns about Lie symmetries are discussed in these textbooks; so we avoid the preliminaries of the Lie symmetries. This chapter discusses the application of the Lie symmetry method and conservation laws for some integer order differential equations. However, some new and different approaches such as the Nucci's method [143, 89, 13, 129] are investigated. Among analytical methods for differential equations, the invariant subspace method is a very close one to the invariance theory, which plays an important role in the Lie symmetry analysis. We refer the interested readers to this topic in [60, 166, 80, 40, 12, 65, 122].

1.1 Classical Lie symmetry analysis

Various types of Lie symmetry method have been introduced up to now, e.g., classical [63, 120, 123, 42, 189, 153, 159, 158, 39, 91], nonclassical [32, 140, 139, 88] and approximate [58, 46, 93] Lie symmetries. Moreover, there are some numerical methods which are based upon Lie groups [76, 168, 10, 9, 82, 73, 4, 2, 70, 78, 79, 86, 74]. Briefly, a symmetry of a differential equation is a transformation which maps every solution of the differential equation to another solution of the same equation.

Here, we present some preliminaries of Lie Groups and Transformation Groups. The main ingredients for this section are the algebraic concept of a group and the differential-geometric notion of a smooth manifold. The term smooth constrains the overlap functions of any coordinate chart to be C^{∞}

functions. The following definition is the foundation of Lie symmetry methods for differential equations.

Definition 1 *(Lie group) A set G is called a Lie Group if there is given a structure on G satisfying the following three axioms.*
i) G is a group.
ii) G is a smooth manifold.
iii) The group operations

$$\underset{(g,h)\to g.h}{G\times G\to G}, \quad \underset{g\to g^{-1}}{G\to G},$$

are smooth functions.

When the dimension of G is r, we call this group an r-parameter Lie group.

Definition 2 *(Lie Transformation Groups) Let $\mathcal{M}$ be a n-dimensional smooth manifold and G a Lie group. An action T of the group G on $\mathcal{M}$ is a smooth mapping*

$$\underset{T(g,x)\equiv gx\to\bar{x}}{T:G\times\mathcal{M}\to\mathcal{M}}$$

with the following properties:

$$T(e,x)=x,\ T(a,T(b,x))=T(ab,x)$$

for any $x\in\mathcal{M},\ g,a,b\in G,\ e\in G$ the unit element. Then G is called a Lie transformation group of the manifold $\mathcal{M}$.

It is well known that the applications of symmetry groups to differential equations include:

- mapping solutions to other solutions
- integration of ordinary differential equations in formulas
- constructing invariant (similarity) solutions, that is, solutions which are invariant under the action of a subgroup of the admitted group
- detection of linearizing transformations.

To carry out any of these, a true technique for finding symmetries of differential equations is needed. As a general idea, one could insert an arbitrary change of variables into the equation and then impose the new variables to satisfy the same differential equation. This earns a number of differential equations (determining equations) to be satisfied by the transformation. This direct approach is too drastic to be of much use: determining equations may be derived, but solving such a large system of nonlinear equations is usually out of the question. The crucial understanding of Lie was that this problem could prevail by considering the 'infinitesimal' action of the group. In order to define the infinitesimals, we defined a one-parameter Lie group of the form

$$\bar{x} = F(x;\epsilon), \tag{1.1}$$

where ϵ is the group parameter, which, without loss of generality, will be assumed to be defined in such a way that the identity element $\epsilon_0 = 0$. Hence

$$x = F(x;\epsilon)|_{\epsilon=0}. \tag{1.2}$$

Definition 3 *(Infinitesimal Transformation) Given a one parameter Lie group of transformation (1.1), we expand $\bar{x} = F(x;\epsilon)$ into its Taylor series in the parameter ϵ in a neighborhood of $\epsilon = 0$. Then, making use of the fact (1.2), we obtain what is called the infinitesimal transformations of the Lie group of transformation (1.1):*

$$\bar{x} = x + \epsilon\xi(x) + O(\epsilon^2), \tag{1.3}$$

where

$$\xi(x) = \frac{\partial \bar{x}}{\partial \epsilon}|_{\epsilon=0}. \tag{1.4}$$

The components of the vector $\xi(x) = (\xi_1(x), \xi_2(x), \ldots, \xi_n(x))$ are called the infinitesimals of (1.1).

Definition 4 *(Infinitesimal generator) The operator*

$$V = \sum_{i=1}^{n} \xi_i(x)\frac{\partial}{\partial x_i} \tag{1.5}$$

is called the infinitesimal generator (operator) of the one-parameter Lie group of transformations (1.1), where $x = (x_1, x_2, \ldots, x_n) \in \mathbb{R}^n$ and $\xi(x) = (\xi_1(x), \xi_2(x), \ldots, \xi_n(x))$ are the infinitesimals of (1.1)

Besides, each constant in a one-parameter Lie group of transformations leads to a symmetry generator (which is a linear operator). These symmetry generators belong to a one-dimensional linear vector space in which any linear combination of generators is also a linear operator and the way of ordering generators is not important, that is, the symmetry group of transformation commutes, and this leads to the additional structure in the mentioned vector space called the commutator.

Definition 5 *Let G be the one-parameter Lie group of transformations (1.1) with the symmetry generators V_i, $i = 1, 2, \ldots, r$ given by (1.5). The commutator (Lie bracket) $[.,.]$ of two symmetry generators V_i, V_j is the first order operator generated as follows*

$$[V_i, V_j] = V_iV_j - V_jV_i.$$

Definition 6 *(Lie algebra) A Lie algebra $\mathcal{L}$ is a vector space over a field F with a given bilinear commutation law (the commutator) satisfying the properties*

1. *Closure:*
   For $X, Y \in \mathcal{L}$ it follows that $[X, Y] \in \mathcal{L}$.

2. *Bilinearity:*
   $[X, \alpha Y + \beta Z] = \alpha[X, Y] + \beta[X, Z], \quad \alpha, \beta \in F, \quad X, Y, Z \in \mathcal{L}.$

3. *Skew-symmetry:*
   $[X, Y] = -[Y, X].$

4. *Jacobi identity:*
   $[X, [Y, Z]] + [Y, [Z, X]] + [Z, [X, Y]] = 0.$

Now, after brief preliminaries of the Lie symmetry method, we illustrate this technique by different integer order differential equations.

1.1.1 Lie symmetries of the Fornberg-Whitham equation

The Fornberg-Whitham equation (FWE)[53, 84],

$$u_t - u_{xxt} + u_x + uu_x = 3u_x u_{xx} + uu_{xxx} \ , \tag{1.6}$$

has appeared in the study of qualitative behaviors of wave breaking, which is a nonlinear dispersive wave equation. In 1978, Fornberg and Whitham obtained a peaked solution of the form $u(x,t) = A\exp\{\frac{-1}{2}|x - \frac{4}{3}t|\}$, where A is an arbitrary constant. Zhou et al. [190] have found the implicit form of a type of traveling wave solution called kink-like wave solutions and antikink-like wave solutions. After that, they found the explicit expressions for the exact traveling wave solutions, peakons and periodic cusp wave solutions for the FWE [191]. Tian et al. [173], under the periodic boundary conditions, have studied the global existence of solutions to the viscous FWE in L^2. The limit behavior of all periodic solutions as the parameters trend to some special values was studied in [186]. F. Abidi et al. [5] have successfully applied the homotopy analysis method to obtain the approximate solution of FWE and compared those to the solutions given by Adomian decomposition method.
The symmetry groups of the FWE will be generated by the vector field of the form

$$X = \xi^1(t,x,u)\frac{\partial}{\partial t} + \xi^2(t,x,u)\frac{\partial}{\partial x} + \phi(t,x,u)\frac{\partial}{\partial u}. \tag{1.7}$$

The result shows that the symmetry of Eq. (1.6) is expressed by a finite three-dimensional point group containing translation in the independent variables and scaling transformations. The group parameters are denoted by k_i for $i = 1, 2, 3$ and characterize the symmetry of equation. Actually, we find that

Eq. (1.6) admits a three-dimensional Lie algebra $\mathcal{L}_3$ of its classical infinitesimal point symmetries generated by the following vector fields:

$$X_1 = \frac{\partial}{\partial t}, \quad X_2 = \frac{\partial}{\partial x}, \quad X_3 = t\frac{\partial}{\partial x} + \frac{\partial}{\partial u}.$$

Obviously, the Lie algebra of (1.6) is solvable and from the adjoint representation of the symmetry Lie algebra the optimal system of one-dimensional subalgebras corresponds to (1.6) which can be expressed by

$$X_1, \quad X_2, \quad \alpha X_1 + X_3,$$

where $\alpha \in \{-1, 0, 1\}$.

1.1.1.1 Similarity reductions and exact solutions

In order to reduce PDE (1.6) to a system of ODEs with one independent variable, we construct similarity variables and similarity forms of field variables. Using a straightforward analysis, the characteristic equations used to find similarity variables are:

$$\frac{dt}{\xi^1} = \frac{dx}{\xi^2} = \frac{du}{\phi}. \tag{1.8}$$

Integration of first order differential equations corresponding to pairs of equations involving only independent variables of (1.8) leads to similarity variables. We distinguish four cases:
Case 1: For the generator X_1, we have:

$$u(t,x) = S(\zeta), \qquad \zeta = x,$$

where $S(\zeta)$ satisfies the following ODE:

$$S' + SS' - 3S'S'' - SS^{(3)} = 0, \tag{1.9}$$

that admits the only Lie symmetry operator $\frac{\partial}{\partial \zeta}$. Instead of using the usual method based on invariants we apply a more direct method, namely the reduction method [143, 142, 145, 146, 128, 89]. Obtaining the first integrals of ODEs is often sophisticated work as shown in [137]. However, using the mentioned reduction method, the first integrals of the reduced ODEs are easily obtained. Equation (1.9) can be written as an autonomous system of three ODEs of first order, i.e.,

$$\begin{cases} w_1' = w_2, \\ w_2' = w_3, \\ w_3' = \dfrac{w_2 + w_1 w_2 - 3w_2 w_3}{w_1}, \end{cases} \tag{1.10}$$

using the obvious change of dependent variables

$$w_1(\zeta) = S(\zeta), \quad w_2(\zeta) = S'(\zeta), \quad w_3(\zeta) = S''(\zeta).$$

Since this system is autonomous, we can choose w_1 as a new independent variable. Then system (1.10) becomes the following nonautonomous system of two ODEs of first order:

$$\begin{cases} \dfrac{dw_2}{dw_1} = \dfrac{w_3}{w_2}, \\ \dfrac{dw_3}{dw_1} = \dfrac{1 + w_1 - 3w_3}{w_1}. \end{cases} \tag{1.11}$$

We can integrate from the second equation:

$$w_3 = \frac{12a_1 + 3w_1^4 + 4w_1^3}{12w_1^3}, \tag{1.12}$$

with a_1 an arbitrary constant. This solution obviously corresponds to the following first integral of equation (1.9):

$$\frac{S(\zeta)^3}{12}\big(12S''(\zeta) - 3S(\zeta) - 4\big) = a_1.$$

Substituting (1.12) into (1.11) yields

$$\frac{dw_2}{dw_1} = \frac{12a_1 + 3w_1^4 + 4w_1^3}{12w_1^3 w_2};$$

that is a separable first order equation too. Therefore, the general solution is

$$w_2 = \sqrt{\frac{-12a_1 + 12a_2 w_1^2 + 3w_1^4 + 8w_1^3}{18w_1^2}}, \tag{1.13}$$

with a_2 an arbitrary constant. Replacing a_1 into this expression in terms of the original variables S and ζ yields another first integral of equation (1.9):

$$\frac{2S(\zeta)S''(\zeta) + 2\left(S'(\zeta)\right)^2 - S^2(\zeta) - 2S(\zeta)}{2} = a_2.$$

Finally, we replace (1.13) from (1.10) into the first equation of system (1.10) that becomes the following separable first-order equation

$$r_1' = p\sqrt{\frac{-2a_1 + a_2(p + q - 2a_1)r_1 - (p + q)r_1^2}{pr_1}},$$

and its general solution is implicitly expressed by

$$\int \sqrt{\frac{18w_1^2}{-12a_1 + 12a_2 w_1^2 + 3w_1^4 + 8w_1^3}}\,dw_1 = \zeta + a_3,$$

and replacing w_1 with $S(\zeta)$ yields the general solution of (1.9).

An explicit subclass of solutions can be obtained if one assumes $a_1 = 0$. Thus

$$u(t,x) = \frac{16 - 36a_2 + e^{\pm(x+a_3)} - 8e^{\pm\left(\frac{x+a_3}{2}\right)}}{6e^{\pm\left(\frac{x+a_3}{2}\right)}}.$$

Case 2: The solution obtained from generator X_2 is trivial. Thus, we find the traveling wave solution which is achievable from generator $X_1 + X_2$. The similarity variable related to $X_1 + X_2$ is

$$u(t,x) = S(\zeta), \quad \zeta = x - t,$$

where $S(\zeta)$ satisfies the following equation:

$$(1 - S)\, S''' + SS' - 3S'S'' = 0. \tag{1.14}$$

Eq. (1.14) admits the only generator $\frac{\partial}{\partial\zeta}$. Therefore it is not possible to solve it by current Lie symmetry methods and we apply the reduction method. This equation transforms into the following autonomous system of first order equations, i.e.,

$$\begin{cases} w_1' = w_2, \\ w_2' = w_3, \\ w_3' = \dfrac{(3w_3 - w_1)w_2}{1 - w_1}, \end{cases} \tag{1.15}$$

by the change of dependent variables

$$w_1(\zeta) = S(\zeta), \quad w_2(\zeta) = S'(\zeta), \quad w_3(\zeta) = S''(\zeta).$$

Similar to Case 1, let us choose w_1 as the new independent variable. Then (1.15) yields:

$$\begin{cases} \dfrac{dw_2}{dw_1} = \dfrac{w_3}{w_2}, \\ \dfrac{dw_3}{dw_1} = \dfrac{(3w_3 - w_1)}{1 - w_1}. \end{cases} \tag{1.16}$$

The second equation of (1.16) is linear and therefore we have

$$w_3 = \frac{12a_1 + 3w_1^4 - 8w_1^8 + 6w_1^2}{12w_1^3 - 36w_1^2 + 36w_1 - 12}, \tag{1.17}$$

and substituting in the other equation of (1.16) yields:

$$\frac{dw_2}{dw_1} = \frac{12a_1 + 3w_1^4 - 8w_1^3 + 6w_1^2}{12w_2\left(w_1^3 - 3w_1^2 + 3w_1 - 1\right)}. \tag{1.18}$$

Replacing a_1 into this expression in terms of the original variables S and ζ yields a first integral of equation (1.14) as:

$$\frac{S''\left(12S^3 - 36S^2 + 36S - 12\right) - 3S^4 + 8S^3 - 6S^2}{12} = a_1.$$

Eq. (1.18) is separable and the solution is given by

$$w_2 = \sqrt{\frac{-12a_1 + 12a_2w_1^2 - 24a_2w_1 + 12a_2 + 3w_1^4 - 4w_1^3 - w_1^2 + 2w_1 - 1}{12\left(w_1 - 1\right)^2}}, \tag{1.19}$$

where a_2 is another first integral of equation (1.14) as following:

$$SS'' - S'' + (S')^2 + \frac{1 - 6S^2}{12} = a_2.$$

An implicit solution of Eq. (1.14) can be obtained from substituting (1.19) into the first equation of (1.15) and one time integration. However, in a special case, taking $a_1 = 0$ and $a_2 = \frac{1}{12}$ we have

$$\frac{S\left[\sqrt{3}(4 - 3S) + 4\sqrt{S(3S - 4)}\ln\left(6\sqrt{S} + 2\sqrt{9S - 12}\right)\right]}{\sqrt{S^3(3S - 4)}} = \zeta + a_3.$$

Back substitution of variables yields another solution of the Eq. (1.6).

Case 3: For the linear combination $X = \alpha X_1 + X_3$, we are just able to find the invariant solution with respect to $\alpha = 0$. Similarity variables of X_3 are:

$$u(t, x) = \frac{x}{t} + S(\zeta), \quad \zeta = t, \tag{1.20}$$

where $S(\zeta)$ admits the following equation:

$$\zeta S' + S + 1 = 0; \tag{1.21}$$

therefore,

$$S(\zeta) = -1 + \frac{c}{\zeta};$$

thus, we get

$$u(t, x) = \frac{x - t + c}{t}. \tag{1.22}$$

1.1.2 Lie symmetries of the modified generalized Vakhnenko equation

Now, we apply the Lie group analysis to the so-called modified generalized Vakhnenko equation (mGVE) [89]:

$$\frac{\partial}{\partial x}(\mathfrak{D}^2 u + \frac{1}{2}pu^2 + \beta u) + q\mathfrak{D}u = 0, \qquad \left(\mathfrak{D} := \frac{\partial}{\partial t} + u\frac{\partial}{\partial x}\right), \tag{1.23}$$

where p, q and β are arbitrary nonzero constants. This equation was introduced by Morrison and Parkes in 2003 [135]. There the N-soliton solution of the mGVE[1] was found if $p = 2q$.

[1] Actually Morrison and Parkes introduced equation (1.23) but they named it mGVE in the case $p = 2q$ only.

TABLE 1.1: Commutator table of Eq. (1.23)

$[\cdot,\cdot]$	V_1	V_2	V_3
V_1	0	0	$pV_1 - 2\beta V_2$
V_2	0	0	$-pV_2$
V_3	$-pV_1 + 2\beta V_2$	pV_2	0

The Vakhnenko equation (VE) can be obtained from the mGVE if $p = q = 1$ and $\beta = 0$, while the generalized Vakhnenko equation (GVE) corresponds to $p = q = 1$ and β arbitrary. We consider a one-parameter Lie group of infinitesimal transformations with independent variables t, x and dependent variable u

$$\begin{cases} t^* = t + \epsilon\xi^1(t,x,u) + O(\epsilon^2), \\ x^* = x + \epsilon\xi^2(t,x,u) + O(\epsilon^2), \\ u^* = u + \epsilon\phi(t,x,u) + O(\epsilon^2), \end{cases}$$

where ϵ is the group parameter. The associated Lie algebra of infinitesimal symmetries is the set of vector fields of the form

$$V = \xi^1(t,x,u)\frac{\partial}{\partial t} + \xi^2(t,x,u)\frac{\partial}{\partial x} + \phi(t,x,u)\frac{\partial}{\partial u}. \tag{1.24}$$

If $\mathrm{Pr}^{(3)}V$ denotes the third prolongation of V then the invariance condition

$$\mathrm{Pr}^{(3)}V(F)|_{F=0} = 0,$$

with

$$F := \frac{\partial}{\partial x}\left(\mathfrak{D}^2 u + \frac{1}{2}pu^2 + \beta u\right) + q\mathfrak{D}u,$$

yields an overdetermined system of linear PDEs in ξ^1, ξ^2 and ϕ. We found that equation (1.23) admits a three-dimensional Lie symmetry algebra $\mathcal{L}$ spanned by the following generators:

$$V_1 = \frac{\partial}{\partial t},\quad V_2 = \frac{\partial}{\partial x},\quad V_3 = pt\frac{\partial}{\partial t} - (px + 2\beta t)\frac{\partial}{\partial x} - (2\beta + 2pu)\frac{\partial}{\partial u}$$

with commutator Table 1.1. This algebra corresponds[2] to $A_{3,4}$ in the classification by Patera and Winternitz [155] with the following identification:

$$e_1 = -\frac{\beta}{2}V_2,\quad e_2 = pV_1 - \beta V_2,\quad e_3 = -\frac{1}{p}V_3.$$

[2]Although as early as 1897 Bianchi [27] gave the classification of all solvable and nonsolvable real algebras of vector fields in the real space as it was stressed in [35], and in 1963 [138] Mubarakzjanov classified the subalgebras of solvable three- and four-dimensional Lie algebras – a paper that regretfully has never been translated into English – we related our notation to a widely available and more cited paper by Patera and Winternitz [155], that also contains the classification of the subalgebras of nonsolvable three- and four-dimensional Lie algebras.

Then the nonzero conjugacy classes of its subalgebras are:
• two-dimensional

$$\mathcal{L}_{1,2} = \langle e_1, e_3 \rangle, \quad \mathcal{L}_{2,2} = \langle e_2, e_3 \rangle, \quad \mathcal{L}_{3,2} = \langle e_1, e_2 \rangle.$$

• one-dimensional

$$\mathcal{L}_{1,1} = \langle e_1 \rangle, \quad \mathcal{L}_{2,1} = \langle e_2 \rangle, \quad \mathcal{L}_{3,1} = \langle e_3 \rangle, \quad \mathcal{L}_{4,1}(\epsilon) = \langle e_1 + \epsilon e_2 \rangle, \quad \text{with } \epsilon = \pm 1.$$

Each one-dimensional subalgebras reduce equation (1.23) to an ODE as we show next.

• Subalgebra $\mathcal{L}_{1,1}$

This subalgebra is spanned by

$$e_1 = -\frac{\beta}{2}\frac{\partial}{\partial x}$$

and the corresponding invariant surface condition is $u_x = 0$. Therefore substituting $u(t,x) = F(t)$ in (1.23) yields

$$F'(t) = 0 \Rightarrow u(t,x) = C,$$

namely a trivial solution.

• Subalgebra $\mathcal{L}_{2,1}$

This subalgebra is spanned by

$$e_2 = p\frac{\partial}{\partial t} - \beta\frac{\partial}{\partial x}.$$

The reduced equation for this subalgebra is

$$(\beta + pF)^2 F''' + 2p(\beta + pF)F'F'' + p^2(p+q)FF' + p\beta(p+q)F' = 0, \quad (1.25)$$

with similarity independent variable $\xi = x + \frac{\beta}{p}t$ and similarity dependent variable F such that $u(t,x) = F(\xi)$.
Equation (1.25) admits a two-dimensional Lie symmetry algebra spanned by the following operators

$$X_1 = \frac{\partial}{\partial \xi}, \quad X_2 = \xi\frac{\partial}{\partial \xi} + 2(F + \frac{\beta}{p})\frac{\partial}{\partial F},$$

and consequently it can be reduced to a first-order ODE. Instead of using the usual method based on invariants we apply a more direct method, namely the Nucci's reduction method [142]. Equation (1.25) can be written as an autonomous system of three ODEs of first order, i.e.,

$$\begin{cases} w_1' = w_2, \\ w_2' = w_3, \\ w_3' = -\dfrac{2pw_2w_3 + p(p+q)w_2}{\beta + pw_1}, \end{cases}$$

by means of the obvious change of dependent variables

$$w_1(\xi) = F(\xi), \quad w_2(\xi) = F'(\xi), \quad w_3(\xi) = F''(\xi).$$

In order to simplify the third equation we introduce the new dependent variable $r_1(\xi) = \beta + pw_1(\xi)$ such that the system becomes

$$\begin{cases} r_1' = pw_2, \\ w_2' = w_3, \\ w_3' = -\dfrac{2pw_2w_3 + p(p+q)w_2}{r_1}. \end{cases} \tag{1.26}$$

Since this system is autonomous we can choose r_1 as a new independent variable. Then system (1.26) becomes the following nonautonomous system of two ODEs of first order:

$$\begin{cases} \dfrac{dw_2}{dr_1} = \dfrac{w_3}{pw_2}, \\ \dfrac{dw_3}{dr_1} = -\dfrac{p+q+2w_3}{r_1}. \end{cases} \tag{1.27}$$

The second equation is independent from the first equation in (1.27) and also separable: therefore we can integrate it, i.e.,

$$w_3 = \frac{a_1}{r_1^2} - (p+q), \tag{1.28}$$

with a_1 an arbitrary constant. This solution obviously corresponds to the following first integral of equation (1.25):

$$\xi^2\big(F''(\xi) + \frac{p+q}{2}\big) = a_1.$$

Substituting (1.28) into (1.27) yields

$$\frac{dw_2}{dr_1} = \frac{2a_1 - (p+q)r_1^2}{2pr_1^2 w_2}.$$

That is a separable first-order equation too and therefore the general solution is

$$w_2 = \sqrt{\frac{-2a_1 + a_2(p+q-2a_1)r_1 - (p+q)r_1^2}{pr_1}}, \tag{1.29}$$

with a_2 an arbitrary constant. Replacing a_1 into this expression in terms of the original variables F and ξ yields another first integral of equation (1.25):

$$\frac{2\xi^2 F''(\xi) + p\big(\beta + pF(\xi)\big)F'^2(\xi) + (p+q)\big(p^2F^2(\xi) + 2p\beta F(\xi) + \xi^2 + \beta^2\big)}{\big(\beta + pF(\xi)\big)\big(2\xi^2 F''(\xi) + (p+q)\xi^2 - (p+q)\big)} = a_2.$$

Lastly, we replace (1.29) from (1.26) into the first equation of system (1.26) that becomes the following separable first-order equation

$$r_1' = p\sqrt{\frac{-2a_1 + a_2(p+q-2a_1)r_1 - (p+q)r_1^2}{pr_1}},$$

and its general solution is implicitly expressed by[3]

$$\frac{\sqrt{(p+q)(\Theta^2-\Psi^2)r_1}}{(p+q)^2\sqrt{-pr_1\Psi\left(2a_1a_2r_1+2a_1+(p+q)(r_1^2-a_2r_1)\right)}}\times$$
$$\left[2\Psi\times\text{EllipticE}\left(\sqrt{\frac{\Theta+\Psi}{\Lambda+\Psi}},\sqrt{\frac{\Lambda+\Psi}{2\Psi}}\right)+\right.$$
$$\left.(2a_1a_2-(p+q)a_2-\Psi)\times\text{EllipticF}\left(\sqrt{\frac{\Theta+\Psi}{\Lambda+\Psi}},\sqrt{\frac{\Lambda+\Psi}{2\Psi}}\right)\right]=\xi+a_3,$$

where

$$\begin{aligned}\Lambda &= 2a_1a_2-(p+q)a_2,\\ \Theta &= (p+q)(2r_1-a_2)+2a_1a_2,\\ \Psi &= \sqrt{4a_1^2a_2^2-4(p+q)a_1a_2^2+(p+q)^2a_2^2-8(p+q)a_1},\end{aligned}$$

and replacing r_1 with $\beta + pF(\xi)$ yields the general solution of (1.25).

An explicit subclass of solutions can be obtained if one assumes $a_1 = 0$. Thus

$$F(\xi) = \frac{4(a_2-\beta)-p(p+q)(\xi+a_3)^2}{4p},$$

and consequently the following is a class of solutions of (1.23),

$$u(t,x) = \frac{4p(a_2-\beta)-(p+q)(px+pa_3+\beta t)^2}{4p^2}.$$

- Subalgebra $\mathcal{L}_{3,1}$

This subalgebra is spanned by

$$e_3 = -\frac{1}{p}V_3 = -t\frac{\partial}{\partial t} + \left(x + 2\frac{\beta}{p}t\right)\frac{\partial}{\partial x} + 2\left(u+\frac{\beta}{p}\right)\frac{\partial}{\partial u},$$

[3] Here we use the MAPLE elliptic integral notations defined by:

$$\text{EllipticE}(z,k) = \int_0^z \frac{\sqrt{1-k^2t^2}}{\sqrt{1-t^2}}dt,\quad \text{EllipticF}(z,k) = \int_0^z \frac{1}{\sqrt{1-t^2}\sqrt{1-k^2t^2}}dt.$$

and its corresponding invariant surface condition,

$$-tu_t + (x + 2\frac{\beta}{p}t)u_x = 2(u + \frac{\beta}{p}),$$

yields

$$\xi = tx + \frac{\beta}{p}t^2, \quad u(t,x) = \frac{F(\xi)}{t^2} - \frac{\beta}{p}, \tag{1.30}$$

as similarity variables.
Substituting (1.30) into (1.23) gives rise to the following third-order ODE

$$(\xi+F)^2F''' + 2(-\xi+\xi F'+FF')F'' - 2(F')^2 + 2F' + (p+q)FF' + q\xi F' - 2qF = 0, \tag{1.31}$$

that does not possess any Lie point symmetry.
Yet two particular solutions can be found if we assume that F is a second-degree polynomial in ξ. In fact the two solutions are

$$F_1(\xi) = \frac{q}{p+q}\xi - 2\frac{q}{(p+q)^2}, \qquad F_2(\xi) = -\frac{p+q}{4}\xi^2 - \frac{p+q}{p}\xi - \frac{(p+q)^2}{p^2q},$$

which yield the following two solutions of (1.23):

$$u_1(t,x) = \frac{txpq + txq^2 - \beta t^2 q - 2q - \beta t^2 p}{(p+q)^2t^2},$$

and

$$u_2(t,x) = -\frac{(p+q)(\beta^2qt^4 + 2pqx\beta t^3 + 4pqxt + 4(p+q)}{4qp^2t^2} - \frac{x^2p^3 + 8p\beta + qx^2p^2 + 4q\beta}{4p^2},$$

respectively.

- Subalgebras $\mathcal{L}_{4,1}$

Since these two subalgebras are spanned by

$$e_1 + \epsilon e_2 = \epsilon p\frac{\partial}{\partial t} - \beta(\frac{1}{2} + \epsilon)\frac{\partial}{\partial x}, \qquad (\epsilon = \pm 1),$$

two cases have to be considered:

1. $\epsilon = -1$

In this case the invariant surface condition yields the similarity variables $u(t,x) = F(\xi)$ with $\xi = x + \frac{\beta}{2p}t$ that, substituted into (1.23), generates the following autonomous third-order ODE:

$$(\beta+2pF)^2F''' + 4p(\beta+2pF)F'F'' + 4p^2(p+q)FF' + 2p\beta(2p+q)F' = 0, \tag{1.32}$$

that admits the Lie symmetry operator $\frac{\partial}{\partial \xi}$ only.

We apply the reduction method to the equation (1.32) by considering the following system of three first-order ODEs

$$\begin{cases} w_1' &= w_2, \\ w_2' &= w_3, \\ w_3' &= -\dfrac{4p(\beta+2pw_1)w_2w_3 + 4p^2(p+q)w_1w_2 + 2p\beta(2p+q)w_2}{(\beta+2pw_1)^2}, \end{cases}$$

with

$$w_1(\xi) = F(\xi), \quad w_2(\xi) = F'(\xi), \quad w_3(\xi) = F''(\xi).$$

A simplification can be obtained by introducing the transformation $r_1(\xi) = \beta + 2pw_1(\xi)$ so that

$$\begin{cases} r_1' &= 2pw_2, \\ w_2' &= w_3, \\ w_3' &= -\dfrac{2\beta p^2 w_2 + 2p^2 r_1 w_2 + 2pqr_1w_2 + 4pr_1w_2w_3}{r_1^2}. \end{cases} \quad (1.33)$$

If we choose r_1 as the new independent variable, then (1.33) becomes a system of two nonautonomous first-order ODEs, i.e.,

$$\begin{cases} \dfrac{dw_2}{dr_1} &= \dfrac{w_3}{2pw_2}, \\ \dfrac{dw_3}{dr_1} &= -\dfrac{p\beta + (p+q)r_1 + 2w_3r_1}{r_1^2}. \end{cases} \quad (1.34)$$

The second equation of (1.34) can immediately be solved, i.e.,

$$w_3 = \frac{2a_1 - 2p\beta r_1 - (p+q)r_1^2}{2r_1^2}, \quad (1.35)$$

with a_1 an arbitrary constant. This corresponds to the following first integral of equation (1.32):

$$\left(F''(\xi) + \frac{p+q}{2}\right)(\beta + 2pF(\xi))^2 + p\beta(\beta + 2pF(\xi)) = const.$$

Substituting (1.35) into (1.34) yields:

$$\frac{dw_2}{dr_1} = \frac{2(a_1 - p\beta r_1) - (p+q)r_1^2}{4pr_1^2w_2},$$

that is easily solved to give

$$w_2 = \sqrt{\frac{2p\beta r_1\left(a_2 - \log(r_1)\right) - 2a_1(1 + a_2r_1) + (p+q)(a_2r_1 - r_1^2)}{2pr_1}},$$

with a_2 an arbitrary constant. This corresponds to another first integral of equation (1.32), i.e.,

$$\left(\frac{\beta}{2}+pF(\xi)\right)\left(3p\beta+q\beta+2p(p+q)F(\xi)+\left(\frac{\beta}{2}+pF(\xi)\right)F''(\xi)\right)=const.$$

Finally, replacing w_2 into the first equation of system (1.33) yields the following first-order ODE

$$r_1'=2\sqrt{\frac{p\left((p+q+2p\beta)a_2-(p+q)r_1\right)r_1-2p(1+a_2r_1)a_1-2p^2\beta r_1\log(r_1)}{2r_1}}, \tag{1.36}$$

and therefore the last quadrature

$$\int\frac{\sqrt{2r_1}dr_1}{\sqrt{p\left((p+q+2p\beta)a_2-(p+q)r_1\right)r_1-2p(1+a_2r_1)a_1-2p^2\beta r_1\log(r_1)}}$$
$$=\sqrt{2}\xi+a_3.$$

If we assume $\beta=0$ then an elliptic function is obtained.

2. $\epsilon=1$

In this case the invariant surface condition yields the similarity variables $u(t,x)=F(\xi)$ with $\xi=x+\frac{3\beta}{2p}t$ that, substituted into (1.23), yields the following autonomous third-order ODE:

$$(3\beta+2pF)^2F'''+4p(3\beta+2pF)F'F''+4p^2(p+q)FF'+2p\beta(2p+3q)F'=0, \tag{1.37}$$

that admits the Lie symmetry operator $\frac{\partial}{\partial\xi}$ only.

We apply the reduction method to equation (1.37) by considering the following system of three first-order ODEs

$$\begin{cases} w_1'=w_2,\\ w_2'=w_3,\\ w_3'=-\dfrac{4p(3\beta+2pw_1)w_2w_3+4p^2(p+q)w_1w_2+2p\beta(2p+3q)w_2}{(3\beta+2pw_1)^2}, \end{cases}$$

with

$$w_1(\xi)=F(\xi),\quad w_2(\xi)=F'(\xi),\quad w_3(\xi)=F''(\xi).$$

A simplification can be obtained by introducing the transformation $r_1(\xi)=3\beta+2pw_1(\xi)$ so that

$$\begin{cases} r_1'=2pw_2,\\ w_2'=w_3,\\ w_3'=\dfrac{2\beta p^2w_2-2p(p+q)r_1w_2-4pr_1w_2w_3}{r_1^2}. \end{cases} \tag{1.38}$$

If we choose r_1 as the new independent variable, then (1.38) becomes a system of two nonautonomous first-order ODEs

$$\begin{cases} \dfrac{dw_2}{dr_1} = \dfrac{w_3}{2pw_2}, \\ \dfrac{dw_3}{dr_1} = \dfrac{p\beta - (p+q)r_1 - 2w_3 r_1}{r_1^2}. \end{cases} \tag{1.39}$$

The second equation of (1.39) can immediately be solved, i.e.,

$$w_3 = \frac{2a_1 + 2p\beta r_1 - (p+q)r_1^2}{2r_1^2}, \tag{1.40}$$

with a_1 an arbitrary constant. This corresponds to the following first integral of equation (1.37) as:

$$F''(\xi)(3\beta + 2pF(\xi))^2 + (3\beta + 2pF(\xi))(p\beta + 3q\beta + 2p(p+q)F(\xi)) = const.$$

Substituting (1.40) into (1.39) concludes:

$$\frac{dw_2}{dr_1} = \frac{2(a_1 + p\beta r_1) - (p+q)r_1^2}{4pr_1^2 w_2},$$

which is easily solved to give

$$w_2 = \sqrt{\frac{((p+q-2p\beta)a_2 - (p+q)r_1)\, r_1 - 2a_1(1 + a_2 r_1) + 2p\beta r_1 \log(r_1)}{2pr_1}},$$

with a_2 an arbitrary constant. This corresponds to another first integral of equation (1.37), i.e.,

$$\frac{-2\left(2p(p+q)F - p\beta ln(3\beta + 2pF) + pF'^2 + (2p+3q)\beta + (3\beta + 2pF)F''\right)}{2(3\beta + 2pF)^2 F'' + (p+q)(4p^2F^2 - 1) + 3\beta^2(p+3q) + 2p\beta(1 + 4pF + 6qF)}$$
$$= const.$$

Finally, replacing w_2 into the first equation of system (1.38) yields the following first-order ODE

$$r_1' = 2\sqrt{\frac{p\left((p+q-2p\beta)a_2 - (p+q)r_1\right) r_1 - 2p(1 + a_2 r_1)a_1 + 2p^2\beta r_1 \log(r_1)}{2r_1}},$$

and therefore the last quadrature

$$\int \frac{\sqrt{r_1}dr_1}{\sqrt{((p+q-3p\beta)a_2 - (p+q)r_1)\, r_1 - 2(1 + a_2 r_1)a_1 + 2p\beta r_1 \log(r_1)}}$$
$$= \sqrt{2p}\xi + a_3,$$

which cannot be explicitly evaluated. If we assume $\beta = 0$, then an elliptic function is obtained, i.e.,

$$\frac{\sqrt{(p+q)(\Theta^2-\Psi^2)r_1}}{(p+q)^2\sqrt{-pr_1\Psi\left(2a_1a_2r_1+2a_1+(p+q)(r_1^2-a_2r_1)\right)}}\times$$

$$\left[2\Psi\times \text{EllipticE}\left(\sqrt{\frac{\Theta+\Psi}{\Lambda+\Psi}},\sqrt{\frac{\Lambda+\Psi}{2\Psi}}\right)+\right.$$

$$\left.(2a_1a_2-(p+q)a_2-\Psi)\times \text{EllipticF}\left(\sqrt{\frac{\Theta+\Psi}{\Lambda+\Psi}},\sqrt{\frac{\Lambda+\Psi}{2\Psi}}\right)\right] = -\frac{4}{\sqrt{2}}\xi + a_3,$$

where

$$\Lambda = 2a_1a_2 - (p+q)a_2,$$
$$\Theta = (p+q)(2r_1 - a_2) + 2a_1a_2,$$
$$\Psi = \sqrt{4a_1^2a_2^2 - 4(p+q)a_1a_2^2 + (p+q)^2a_2^2 - 8(p+q)a_1}\ .$$

1.1.3 Lie symmetries of the Magneto-electro-elastic circular rod equation

The nonlinear Magneto-electro-elastic circular rod equation (MEE) circular rod [183, 87] reads:

$$u_{tt} - c^2u_{xx} - \frac{c^2}{2}u_{xx}^2 - Nu_{ttxx} = 0, \tag{1.41}$$

where c is the linear longitudinal wave velocity for a MEE circular rod and N is the dispersion variable, both turning on the substance features as well as the geometry of the rod. It is also assumed that the infinite homogeneous MEE circular rod is made of composite BaTiO3-CoFe2O4 with different volume fractions of BaTiO3. The rod has a radius $R = 0.05m$. The material features of the composite are approximated using the simple rule of mixture according to the volume fraction. Some authors acquired traveling wave and solitary wave solutions by using the different methods [105, 187].

Let us to consider a one-parameter Lie group of infinitesimal transformations (1.1.2) and associated vector field of the form (1.24).
If $Pr^{(4)}V$ denotes the fourth prolongation of V then the invariance condition is

$$Pr^{(4)}V(\Delta)|_{\Delta=0} = 0,$$

where $\Delta := u_{tt} - c^2u_{xx} - \frac{c^2}{2}u_{xx}^2 - Nu_{ttxx}$, and yields the following determining equations, which, from solving these equations, we get

$$\xi^1 = C_2 + tC_7,\ \ \xi^2 = C_1,\ \ \phi = C_3 + xC_4 + tC_5 + xtC_6 - (c^2t^2 + 2u + x^2)C_7,$$

where C_i, $(i = 1,\ldots,7)$ are arbitrary constants. Thus the Lie symmetry algebra admitted by Eq. (1.41) is spanned by the following five infinitesimal generators

$$V_1 = \frac{\partial}{\partial x},\ V_2 = \frac{\partial}{\partial t},\ V_3 = \frac{\partial}{\partial u},\ V_4 = x\frac{\partial}{\partial u},\ V_5 = t\frac{\partial}{\partial u},$$
$$V_6 = tx\frac{\partial}{\partial u},\ V_7 = t\frac{\partial}{\partial t} - (c^2t^2 + 2u + x^2)\frac{\partial}{\partial u}. \tag{1.42}$$

Reduction 1. Similarity variables related to $cV_1 + V_2 = c\frac{\partial}{\partial x} + \frac{\partial}{\partial t}$ are $u(t,x) = F(\xi)$, where $\xi = x - ct$ and satisfies the following equation:

$$(F''(\xi))^2 + 2NF^{(4)}(\xi) = 0. \tag{1.43}$$

To find the exact solutions of Eq. (1.43), we first use $G(\xi) = F''(\xi)$ to reduce it in the following form:

$$G^2(\xi) + 2NG''(\xi) = 0. \tag{1.44}$$

Here we want to apply the Nucci's reduction method to Eq. (1.44). More details of the reduction method can be found in [146, 128, 89, 84]. This equation transforms into the following autonomous system of first order, i.e.,

$$\begin{cases} w_1' = w_2, \\ w_2' = -\dfrac{w_1^2}{2N}, \end{cases} \tag{1.45}$$

by the change of dependent variables

$$w_1(\xi) = G(\xi),\quad w_2(\xi) = G'(\xi).$$

Let us choose w_1 as the new independent variable. Then (1.45) yields:

$$\frac{dw_2}{dw_1} = -\frac{w_1^2}{2Nw_2}. \tag{1.46}$$

From one time integration of (1.46) we have

$$w_2 = \sqrt{\frac{6Na_1 - w_1^3}{3N}}. \tag{1.47}$$

In this step we return the w_1 as a dependent variable, i.e., $w_1 = w_1(\xi)$. Therefore by putting (1.47) in (1.45) we have a separable first order equation

$$w_1' = \sqrt{\frac{6Na_1 - w_1^3}{3N}},$$

from which its implicit solution is easily obtainable. In a special case, by supposition $a_1 = 0$ an explicit solution can be obtained as follows:

$$w_1(\xi) = -\frac{12N}{\xi^2 + 2\sqrt{3}a_2\xi + 3a_2^2},$$

or

$$F''(\xi) = -\frac{12N}{\xi^2 + 2\sqrt{3}a_2\xi + 3a_2^2},$$

which leads to

$$F(\xi) = a_3 + a_4\xi + 12N\ln(3a_2 + \sqrt{3}\xi),$$

and hence

$$u(t,x) = a_3 + a_4(x - ct) + 12N\ln\left(3a_2 + \sqrt{3}(x - ct)\right).$$

Reduction 2. In this case we consider the generator

$$V_1 + V_2 + V_3 = \frac{\partial}{\partial t} + \frac{\partial}{\partial x} + \frac{\partial}{\partial u}, \tag{1.48}$$

with corresponding similarity variables:

$$u(t,x) = t + F(\xi), \quad \xi = x - t. \tag{1.49}$$

After substituting the similarity variables into (1.41) we find

$$2(1 - c^2)F''(\xi) - c^2(F''(\xi))^2 - 2NF^{(4)}(\xi) = 0.$$

Similar to the previous case, after substituting $F''(\xi) = G(\xi)$ in the above ODE, we obtain

$$2(1 - c^2)G(\xi) - c^2G^2(\xi) - 2NG''(\xi) = 0. \tag{1.50}$$

Equation (1.50) transforms into the following autonomous system of first order, namely

$$\begin{cases} w_1' = w_2, \\ w_2' = \dfrac{2(1 - c^2)w_1 - c^2w_1^2}{2N}, \end{cases} \tag{1.51}$$

by the change of dependent variables

$$w_1(\xi) = G(\xi), \quad w_2(\xi) = G'(\xi).$$

Let us choose w_1 as the new independent variable. Then (1.51) yields:

$$\frac{dw_2}{dw_1} = \frac{2(1 - c^2)w_1 - c^2w_1^2}{2Nw_2}. \tag{1.52}$$

From one time integration of (1.52) we have

$$w_2 = \pm\sqrt{\frac{3w_1^2 - 3c^2w_1^2 + 6Na_1 - c^2w_1^3}{3N}}. \tag{1.53}$$

In this step we return the w_1 as a dependent variable, i.e., $w_1 = w_1(\xi)$. Therefore by putting (1.53) in (1.51) we have a separable first order equation

$$w_1' = \pm\sqrt{\frac{3w_1^2 - 3c^2w_1^2 + 6Na_1 - c^2w_1^3}{3N}},$$

from which its implicit solution is easily obtainable. In a special case, by supposition $a_1 = 0$ an explicit solution can be obtained as follows:

$$w_1(\xi) = 3\left(1 - \frac{1}{c^2}\right)\left(-1 - \tan\left[\frac{1}{2}\left(-\sqrt{\frac{c^2-1}{N}}\xi - \sqrt{3\,(c^2-1)}a_2\right)\right]^2\right),$$

or

$$F''(\xi) = 3\left(1 - \frac{1}{c^2}\right)\left(-1 - \tan\left[\frac{1}{2}\left(-\sqrt{\frac{c^2-1}{N}}\xi - \sqrt{3\,(c^2-1)}a_2\right)\right]^2\right),$$

and owing to this fact we obtain

$$F(\xi) = a_3 + a_4\xi + \frac{12N}{c^2}\ln\left[\cos\left(-\sqrt{\frac{c^2-1}{N}}\left(\sqrt{3Na_2} - \xi\right)\right)\right].$$

Hence from (1.49) we obtain a general solution

$$u(t,x) = t + a_3 + a_4(x-t) + \frac{12N}{c^2}\ln\left[\cos\left(-\sqrt{\frac{c^2-1}{N}}\left(\sqrt{3Na_2} - x + t\right)\right)\right].$$

Reduction 3.
Now, we consider the generator

$$V_1 + V_2 + V_4 = \frac{\partial}{\partial t} + \frac{\partial}{\partial x} + x\frac{\partial}{\partial u}. \tag{1.54}$$

Similarity variables related to (1.54) are

$$u(t,x) = \frac{1}{2}\left(-t^2 + 2tx + 2F(\xi)\right), \quad \xi = x - t, \tag{1.55}$$

where

$$2(1-c^2)F''(\xi) - c^2(F''(\xi))^2 - 2\left(1 + NF^{(4)}(\xi)\right) = 0.$$

If we take $F''(\xi) = G(\xi)$ in the above ODE, we obtain

$$2(1-c^2)G(\xi) - c^2G^2(\xi) - 2\left(1 + NG''(\xi)\right) = 0. \tag{1.56}$$

Equation (1.56) transforms into the following autonomous system of first order, i.e.,

$$\begin{cases} w_1' = w_2, \\ w_2' = \dfrac{2(1-c^2)w_1 - c^2w_1^2 - 2}{2N}, \end{cases} \tag{1.57}$$

by the change of dependent variables

$$w_1(\xi) = G(\xi), \quad w_2(\xi) = G'(\xi).$$

Let us choose w_1 as the new independent variable. Then (1.57) yields:

$$\frac{dw_2}{dw_1} = \frac{2(1-c^2)w_1 - c^2w_1^2 - 2}{2Nw_2}. \tag{1.58}$$

One time integration of (1.58) produces

$$w_2 = \pm\sqrt{\frac{3w_1^2 - 3c^2w_1^2 + 6Na_1 - c^2w_1^3}{3N}}. \tag{1.59}$$

In this step we return the w_1 as a dependent variable, i.e., $w_1 = w_1(\xi)$. Therefore by putting (1.59) in (1.57) we have a separable first order equation

$$w_1' = \pm\sqrt{\frac{3w_1^2 - 3c^2w_1^2 + 6Na_1 - c^2w_1^3}{3N}},$$

from which its implicit solution is easily obtainable as follows:

$$\xi - 3\int\sqrt{\frac{N}{-18w_1 + 9\left(1-c^2\right)w_1^2 + 18Na_1 - 3c^2w_1^3}}dw_1 = a_2.$$

Reduction 4.

Now, we consider the generator

$$V_1 + V_5 + V_6 = \frac{\partial}{\partial t} + t\frac{\partial}{\partial u} + tx\frac{\partial}{\partial u}.$$

Corresponding similarity variables are

$$u(t,x) = t\left(x + \frac{x^2}{2}\right) + F(\xi), \quad \xi = t,$$

where

$$-c^2\xi\left(2+\xi\right) + 2F''(\xi) = 0. \tag{1.60}$$

Equation (1.60) is a linear ODE and its solution can be written as follows:

$$F\left(\xi\right) = \frac{1}{4}c^2\xi\left(2+\xi\right)x^2 + a_1x + a_2,$$

or equivalently:

$$u(t,x) = t\left(x + \frac{x^2}{2}\right) + \frac{1}{4}c^2 t\,(2+t)\,x^2 + a_1 x + a_2.$$

Reduction 5.
Suppose the generator

$$V_2 + V_3 + V_4 = \frac{\partial}{\partial t} + \frac{\partial}{\partial u} + x\frac{\partial}{\partial u}. \tag{1.61}$$

Similarity variables related to (1.61) are

$$u(t,x) = t\,(1+x) + F(\xi), \quad \xi = t, \tag{1.62}$$

where

$$c^2 F''(\xi)\,(2 + F''(\xi)) = 0. \tag{1.63}$$

Solving Eq. (1.63) and (1.62) yields the exact solution:

$$u(t,x) = t\,(1+x) - x^2 + a_1 x + a_2.$$

Reduction 6.

Finally, we consider the generator

$$V_1 + V_4 + V_5 + V_6 = \frac{\partial}{\partial x} + (x + t + xt)\,\frac{\partial}{\partial u},$$

with corresponding similarity variables

$$u(t,x) = tx + (1+t)\,\frac{x^2}{2} + F(\xi), \quad \xi = t, \tag{1.64}$$

where

$$-c^2\left(3 + 4\xi + \xi^2\right) + 2F''(\xi) = 0. \tag{1.65}$$

Solving Eq. (1.65) and (1.64) concludes the exact solution:

$$u(t,x) = tx + (1+t)\,\frac{x^2}{2} + \frac{c^2}{4}\left(3 + 4t + t^2\right)x^2 + a_1 x + a_2.$$

1.1.4 Lie symmetries of the couple stress fluid-filled thin elastic tubes

In most theoretical investigations on arterial pulse wave transmission through a thin elastic walled tube, blood thickening due to rise in red blood cells has been assumed to be insignificant [171]. Many researchers have discussed and presented new models about flow in fluid-filled thin elastic tubes. Adesanya and co-workers have presented an investigation about the equations governing the fluid flow. They have used some assumptions and variable transformations to reduce the fluid flow equation to a new style of an evolution equation [6, 85]:

$$u_\tau + a_1 u u_\xi - a_2 u_{\xi\xi} + a_3 u_{\xi\xi\xi} + a_4 u_{\xi\xi\xi\xi} = 0. \tag{1.66}$$

Equation (1.66) in the limiting case as $a_2, a_3, a_4 \to 0$ gives the inviscid Burger's equation. Also, $a_3, a_4 \to 0$ gives the viscous Burger's equation, $a_4 \to 0$ is the KdV-Burger's equation, while $a_3 = 0$, $a_2 = -1$ gives the Kuramoto-Sivashinsky (KS) equation.

Let us consider a one-parameter Lie group of infinitesimal transformations:

$$\begin{cases} \tau^* = \tau + \epsilon\zeta^1(\tau, \xi, u) + O(\epsilon^2), \\ \xi^* = \xi + \epsilon\zeta^2(\tau, \xi, u) + O(\epsilon^2), \\ u^* = u + \epsilon\phi(\tau, \xi, u) + O(\epsilon^2), \end{cases}$$

where ϵ is the group parameter. The associated Lie algebra of infinitesimal symmetries is the set of the vector field of the form

$$V = \zeta^1(\tau, \xi, u)\frac{\partial}{\partial \tau} + \zeta^2(\tau, \xi, u)\frac{\partial}{\partial \xi} + \phi(\tau, \xi, u)\frac{\partial}{\partial u}.$$

If $Pr^{(4)}V$ denotes the fourth prolongation of V then the invariance condition is

$$Pr^{(4)}V(\Delta)|_{\Delta=0} = 0,$$

where $\Delta := u_\tau + a_1 u u_\xi - a_2 u_{\xi\xi} + a_3 u_{\xi\xi\xi} + a_4 u_{\xi\xi\xi\xi}$, and yields an overdetermined system of linear PDEs in ζ^1, ζ^2 and ϕ, the so-called determining equations, from which solving these equations in different cases we get:

Case 1: $a_1 a_4 \neq 0, \quad a_3 = 0, \ a_2 = -1.$
This case is related to the KS equation and we have

$$\zeta^1 = C_1, \quad \zeta^2 = a_1 \tau C_2 + C_3, \quad \phi = C_2,$$

where C_1, C_2 and C_3 are arbitrary constants. Thus the Lie symmetry algebra admitted by Eq. (1.66) is spanned by the following three infinitesimal generators

$$V_1 = \frac{\partial}{\partial \tau}, \ V_2 = \frac{\partial}{\partial \xi}, \ V_3 = a_1 \tau \frac{\partial}{\partial \xi} + \frac{\partial}{\partial u}. \tag{1.67}$$

We present below a reduction and related solution with some different generators:

Reduction 1.1. Similarity variables related to $V_1+\nu V_2$ are $u(\tau,\xi)=\Phi(\theta)$, where $\theta=\xi-\nu\tau$ and satisfies the following equation:

$$(a_1\Phi(\theta)-\nu)\,\Phi'(\theta)+\Phi''(\theta)+a_4\Phi^{(4)}(\theta)=0. \tag{1.68}$$

Eq. (1.68) has the following soliton solution

$$\Phi(\theta)=\frac{1}{19a_1}\left[30+19v-\frac{3(30+19v)}{2\,(1+\cosh\theta+\sinh\theta)^2}+\frac{120}{(1+\cosh\theta+\sinh\theta)^3}\right].$$

Case 2: $a_1a_2a_3a_4\neq 0$.
This case is a more general one and we can get

$$\zeta^1=C_2,\quad \zeta^2=a_1\tau C_1+C_3,\quad \phi=C_1,$$

where C_1,C_2 and C_3 are arbitrary constants. Thus, the Lie symmetry algebra admitted by Eq. (1.66) is spanned by the following three infinitesimal generators

$$V_1=\frac{\partial}{\partial\tau},\quad V_2=\frac{\partial}{\partial\xi},\quad V_3=a_1\tau\frac{\partial}{\partial\xi}+\frac{\partial}{\partial u}. \tag{1.69}$$

We present below a reduction and related solutions with some different generators:

Reduction 2.1. Similarity variables related to $V_1+\nu V_2$ are $u(\tau,\xi)=\Phi(\theta)$, where $\theta=\xi-\nu\tau$ and satisfies the following equation:

$$(a_1\Phi(\theta)-\nu)\,\Phi'(\theta)-a_2\Phi''(\theta)+a_3\Phi'''(\theta)+a_4\Phi^{(4)}(\theta)=0. \tag{1.70}$$

Eq. (1.70) has the following soliton solution of Eq. (1.66):
Family 2.1.

$$\Phi(\theta)=\frac{120a_4}{a_1}\left[\frac{1}{2}+\frac{v}{120a_4}-\frac{1}{(1+\cosh\theta+\sinh\theta)^3}\right].$$

Family 2.2.

$$\Phi(\theta)=c\left[\frac{1}{(1+\cosh\theta+\sinh\theta)^3}+\frac{i-3}{2\,(1+\cosh\theta+\sinh\theta)^2}\right.$$
$$\left.+\frac{1-i}{2\,(1+\cosh\theta+\sinh\theta)}-\frac{(48-4i)\,a_4+v}{120a_4}\right].$$

Family 2.3.

$$\Phi(\theta)=c\bigg[\frac{1}{(1+\cosh\theta+\sinh\theta)^3}+\frac{1}{(1+\cosh\theta+\sinh\theta)^2}$$
$$+\frac{1}{(1+\cosh\theta+\sinh\theta)}+\frac{v-45a_4}{120a_4}\bigg].$$

Family 2.4.

$$\Phi(\theta)=c\left[\frac{1}{(1+\cosh\theta+\sinh\theta)^3}-\frac{1}{(1+\cosh\theta+\sinh\theta)^2}-\frac{v-6a_4}{120a_4}\right].$$

Case 3: $a_1=0,\ a_2a_3\neq 0,\ a_4=-\frac{3a_3^2}{8a_2}$.
This case is a more general one and we can get

$$\zeta^1=C_1+\tau C_2,\ \ \zeta^2=\frac{\xi C_2}{4}-\frac{3a_3^3\tau C_2}{64a_4^2}+C_4,\ \ \phi=-\frac{a_3\xi uC_2}{16a_4}+uC_3+g(\tau,\xi),$$

where C_1, C_2, C_3 and C_4 are arbitrary constants and $g(\tau,\xi)$ satisfies the following equation:

$$3a_3^2g_{\xi\xi}+8a_3a_4g_{\xi\xi\xi}+8a_4^2g_{\xi\xi\xi\xi}+8a_4g_\tau=0.$$

Thus the Lie symmetry algebra admitted by Eq. (1.66) is spanned by the following infinitesimal generators

$$V_1=\frac{\partial}{\partial\tau},\ V_2=\frac{\partial}{\partial\xi},\ V_3=u\frac{\partial}{\partial u},$$
$$V_4=64\tau\frac{\partial}{\partial\tau}+\left(16\xi-\frac{3a_3^3\tau}{a_4^2}\right)\frac{\partial}{\partial\xi}-\frac{4a_3\xi u}{a_4}\frac{\partial}{\partial u},\ V_\infty=g(\tau,\xi)\frac{\partial}{\partial u}.$$

We present below a reduction and related solutions with some different generators:

Reduction 3.1. Similarity variables related to $V_1+V_2+V_3$ are $u(\tau,\xi)=e^\tau\Phi(\theta)$, where $\theta=\xi-\tau$ and satisfies the following equation:

$$8a_2\Phi(\theta)-8a_2\Phi'(\theta)-8a_2^2\Phi''(\theta)+8a_2a_3\Phi'''(\theta)-3a_3^2\Phi^{(4)}(\theta)=0. \quad (1.71)$$

The reduced Eq. (1.71) is linear so the solutions can be easily seen.

Case 4: $a_4=0,\ a_1a_2a_3\neq 0$.
This case is corresponding to the KdV-Burger's equation and we have

$$\zeta^1=C_2,\ \ \zeta^2=a_1\tau C_1+C_3,\ \ \phi=C_1,$$

where C_1, C_2 and C_3 are arbitrary constants. Thus, the Lie symmetry algebra admitted by Eq. (1.66) is spanned by the following three infinitesimal generators

$$V_1 = \frac{\partial}{\partial \tau}, \ V_2 = \frac{\partial}{\partial \xi}, \ V_3 = a_1 \tau \frac{\partial}{\partial \xi} + \frac{\partial}{\partial u},$$

We present below a reduction and related solutions with some different generators:

Reduction 4.1. Similarity variables related to $V_1 + \nu V_2$ are $u(\tau, \xi) = \Phi(\theta)$, where $\theta = \xi - \nu\tau$ and satisfy the following equation:

$$(a_1 \Phi(\theta) - \nu) \Phi'(\theta) - a_2 \Phi''(\theta) + a_3 \Phi'''(\theta) = 0. \tag{1.72}$$

Eq. (1.72) has the following soliton solution

$$\Phi(\theta) = \frac{1}{a_1} \left[-\frac{12 a_3}{(1 + \cosh\theta + \sinh\theta)^2} + \frac{24 a_3}{1 + \cosh\theta + \sinh\theta} + v - 6a_3 \right].$$

Case 5: $a_3 = a_4 = 0, \ a_1 a_2 \neq 0.$
This case is corresponding to the viscous Burger's equation and we have

$$\zeta^1 = -\tau^2 C_1 + C_3 + \tau C_4, \quad \zeta^2 = a_1 \tau C_2 + \frac{1}{2}\xi(C_4 - 2\tau C_1) + C_5,$$
$$\phi = -\frac{\xi C_1}{a_1} + \tau u C_1 + C_2 - \frac{u C_4}{2},$$

where C_i, $(i = 1, \ldots, 5)$ are arbitrary constants. Thus, the Lie symmetry algebra admitted by Eq. (1.66) is spanned by the following five infinitesimal generators

$$V_1 = \frac{\partial}{\partial \tau}, \ V_2 = \frac{\partial}{\partial \xi}, \ V_3 = a_1 \tau \frac{\partial}{\partial \xi} + \frac{\partial}{\partial u},$$
$$V_4 = 2\tau \frac{\partial}{\partial \tau} + \xi \frac{\partial}{\partial \xi} - u \frac{\partial}{\partial u}, \ V_5 = -\tau^2 \frac{\partial}{\partial \tau} - \tau\xi \frac{\partial}{\partial \xi} + \left(\tau u - \frac{\xi}{a_1}\right) \frac{\partial}{\partial u}.$$

We list below some reductions and related solutions with some different generators:

Reduction 5.1. Similarity variables related to V_5 are $u(\tau, \xi) = \frac{\xi + a_1 \Phi(\theta)}{a_1 \tau}$, where $\theta = \frac{\xi}{\tau}$ and satisfies the following equation:

$$a_1 \Phi(\theta) \Phi'(\theta) - a_2 \Phi''(\theta) = 0. \tag{1.73}$$

Eq. (1.73) has the following soliton solutions of Eq. (1.66) :
Family 5.1.1.

$$\Phi(\theta) = \frac{a_2}{a_1} \left[1 - \frac{2}{1 + \cosh\theta + \sinh\theta} \right],$$

so the solution of Eq. (1.66) is as follows

$$u(\tau,\xi)=\frac{1}{a_1\tau}\left(\xi+a_2-\frac{2a_2}{1+\cosh\frac{\xi}{\tau}+\sinh\frac{\xi}{\tau}}\right).$$

Family 5.1.2.

$$\Phi(\theta)=c\tanh\left(\frac{-a_1c}{2a_2}\theta\right),$$

so the solution of the Eq. (1.66) is as follows

$$u(\tau,\xi)=\frac{\xi+a_1c\tanh\left(\frac{-a_1c}{2a_2}\left(\frac{\xi}{\tau}\right)\right)}{a_1\tau}.$$

Reduction 5.2. Similarity variables related to $V_1+\nu V_2$ are $u(\tau,\xi)=\Phi(\theta)$, where $\theta=\xi-\nu\tau$ and satisfy the following equation:

$$\left(a_1\Phi(\theta)-\nu\right)\Phi'(\theta)-a_2\Phi''(\theta)=0. \tag{1.74}$$

Eq. (1.74) has the following soliton solutions of Eq. (1.66) :

Family 5.2.1.

$$\Phi(\theta)=-\frac{1}{a_1}\left[a_2+v+\frac{2a_2}{1+\cosh\theta+\sinh\theta}\right].$$

Family 5.2.2.

$$\Phi(\theta)=\frac{1}{a_1}\left[v-c\tanh\left(\frac{a_1c}{2a_2}\theta\right)\right].$$

1.1.5 Lie symmetries of the generalized Kadomtsev-Petviashvili-modified equal width equation

The Kadomtsev-Petviashvili (KP) equation [102] is one of the familiar 2-dimensional generalizations of the KdV equation which describes the evolution of a weakly nonlinear and weakly dispersive wave, where it appears in the form [83]:

$$\left(u_t+auu_x+u_{xxx}\right)_x+u_{yy}=0.$$

Waves in ferromagnetic media and shallow water waves with weakly nonlinear restoring forces and many significant physical phenomena are described by the KP equation.

The generalized form of the modified equal width (MEW) equation [178, 49, 50]:

$$u_t+a\left(u^3\right)_x-bu_{xxt}=0,$$

which appears in many physical applications, is considered by Wazwaz [177] in the KP sense given by

$$(u_t + a(u^n)_x + bu_{xxt})_x + ru_{yy} = 0. \tag{1.75}$$

In this section, we consider the time variable version of the generalized KP-MEW equation (1.75) as follows:

$$(u_t + a(t)(u^n)_x + b(t)u_{xxt})_x + r(t)u_{yy} = 0. \tag{1.76}$$

Let us consider the Lie symmetries of (1.76) for some special cases which pass the Painlevè test.

Case (a): $n = 2$.

The vector field corresponding to the mentioned group has the following form:

$$V = \xi(x,y,t,u)\frac{\partial}{\partial x} + \eta(x,y,t,u)\frac{\partial}{\partial y} + \tau(x,y,t,u)\frac{\partial}{\partial t} + \phi(x,y,t,u)\frac{\partial}{\partial u}. \tag{1.77}$$

Applying the fourth prolongation $Pr^{(4)}(V)$ to Eq. (1.76) with $n = 2$, we find

$$\xi = a_1 x + b_1, \quad \eta = a_3 y + b_3, \quad \tau = p(t), \quad \phi = ua_2 + (yh_1(t) + h_2(t)),$$

where a_1, a_2, a_3, b_1 and b_3 are arbitrary constants and $a(t)$, $b(t)$ and $r(t)$ have the following forms:

$$a(t) = c_3 e^{\int_1^t \frac{a_1 - a_2}{p(s)} ds}, \qquad b(t) = c_2 e^{\int_1^t \frac{2a_1}{p(s)} ds}, \qquad r(t) = c_1 e^{\int_1^t \frac{2a_3 - a_1}{p(s)} ds}$$

$$\begin{cases} a(t)b(t)\Big(yh_1(t) + h_2(t)\Big) = 0, \\ a(t)r(t)\Big(yh_1(t) + h_2(t)\Big) = 0. \end{cases}$$

The infinitesimal generators of the corresponding Lie algebra are given by the following cases.

Case (a.1)

For $a(t), b(t), r(t) \neq 0$ and $yh_1(t) + h_2(t) = 0$, we have

$$\begin{array}{lll} V_1 = x\frac{\partial}{\partial y}, & V_2 = u\frac{\partial}{\partial u}, & V_3 = y\frac{\partial}{\partial t}, \\ V_4 = \frac{\partial}{\partial y}, & V_5 = \frac{\partial}{\partial t}, & V_6 = p(t)\frac{\partial}{\partial x}. \end{array}$$

In the special case, we consider the invariant solution of Eq. (1.76) for the vector field

$$V_2 + V_4 + V_5 + V_6 = \frac{\partial}{\partial y} + \frac{\partial}{\partial t} + p(t)\frac{\partial}{\partial x} + u\frac{\partial}{\partial u},$$

and $p(t) = 1$ for which corresponding characteristic equations are

$$\frac{dx}{1} = \frac{dy}{1} = \frac{dt}{1} = \frac{du}{u}; \tag{1.78}$$

by solving (1.78) we obtain

$$\rho = x - t, \qquad \psi = y - t, \qquad u(x, y, t) = e^{t+H(\rho,\psi)}. \tag{1.79}$$

Substituting (1.79) in (1.76), we have

$$\begin{aligned}
&- e^{t+H(\rho,\psi)}\Big((H_\rho)^2 + H_{\rho\rho} + H_\rho H_\psi + H_{\rho\psi} - H_\rho\Big) + 2a(t)(H_\rho)^2 e^{2t+2H(r,s)}\\
&+ 2a(t)e^{2t+2H(\rho,\psi)}\Big((H_\rho)^2 + H_{\rho\rho}\Big) - b(t)e^{t+H(r,s)}\Big((H_\rho)^4 + 6H_{\rho\rho}(H_\rho)^2\\
&+ (H_\rho)^3 H_\psi + 4H_{\rho\rho\rho}H_\rho + 3(H_{\rho\rho})^2 + 3H_{\rho\rho}H_\rho H_\psi + 3H_{\rho\psi}(H_\rho)^2 - (H_\rho)^3\\
&+ H_{\rho\rho\rho\rho} + H_{\rho\rho\rho}H_\psi + 3H_{\rho\rho\psi}H_\rho + 3H_{\rho\rho}H_{\rho\psi} - 3H_{\rho\rho}H_\rho + H_{\rho\rho\rho\psi} - H_{\rho\rho\rho}\Big)\\
&+ r(t)e^{t+H(\rho,\psi)}\Big((H_\psi)^2 + H_{\psi\psi}\Big) = 0.
\end{aligned}$$

Then we obtain

$$H(\rho, \psi) = \ln\left(\mathcal{F}_1(\rho)\psi + \mathcal{F}_2(\rho)\right),$$

where $\mathcal{F}_1(\rho)$ and $\mathcal{F}_2(\rho)$ are arbitrary functions. Hence, an exact solution of Eq. (1.76), extracted from this case, is:

$$u(x, y, t) = e^t\big[(y - t) \times \mathcal{F}_1(x - t) + \mathcal{F}_2(x - t)\big]. \tag{1.80}$$

Obtained solution (1.80) is a general class of solutions for the Eq. (1.76), which to the best of our knowledge can be obtained only by the Lie symmetry method.

Case (a.2)

For the case $a(t) = 0$, i.e.,

$$\left(u_t + b(t)u_{xxt}\right)_x + r(t)u_{yy} = 0, \tag{1.81}$$

we obtain

$$\begin{aligned}
\xi &= b_1 + a_1 x, \qquad & \tau &= \frac{a_4}{p(t)} - \frac{\int_1^t a_1 p(\tau)d\tau}{p(t)} + \frac{2\int_1^t b_3 p(\tau)d\tau}{p(t)},\\
\eta &= c_3 + b_3 y, \qquad & \phi &= u(d_1 + \frac{1}{2}b_3 + a_1) + g(x, y, t),
\end{aligned}$$

and $a_1, a_4, b_1, b_3, c_3, d_1$ are arbitrary constants. Moreover, in this case $r(t)$ is an arbitrary function, but $b(t)$ has the following form:

$$b(t) = a_5\Big(a_4 + \int_1^t -(a_1 - 2b_3)p(\tau)d\tau\Big)^{\frac{-2a_1}{a_1-2b_3}}.$$

Therefore, we get

$$V_1 = u\frac{\partial}{\partial u} - \Big(\frac{\int_1^t p(\tau)d\tau}{p(t)}\Big)\frac{\partial}{\partial x} + x\frac{\partial}{\partial y},\ V_2 = \frac{\partial}{\partial y},$$
$$V_3 = \frac{u}{2}\frac{\partial}{\partial u} + 2\Big(\frac{\int_1^t p(\tau)d\tau}{p(t)}\Big)\frac{\partial}{\partial x} + y\frac{\partial}{\partial t},\ V_4 = \frac{1}{p(t)}\frac{\partial}{\partial x},$$
$$V_5 = u\frac{\partial}{\partial u},\ V_6 = \frac{\partial}{\partial t},\ V_7 = g(x,y,t)\frac{\partial}{\partial u},$$

where $g(x,y,t)$ is an arbitrary function satisfying Eq. (1.81). From the vector field

$$V_2 + V_4 + V_5 + V_6 = \frac{\partial}{\partial y} + \frac{1}{p(t)}\frac{\partial}{\partial x} + u\frac{\partial}{\partial u} + \frac{\partial}{\partial t},$$

we obtain the similarity variables as follows:

$$\rho = y - t, \qquad \psi = x - \int \frac{dt}{p(t)}, \qquad u = e^{t+H(\rho,\psi)}. \tag{1.82}$$

Substituting (1.82) in (1.76) yields

$$\begin{aligned}
&(H_\rho)^3 H_\psi b(t)p(t) + (H_\rho)^4 b(t) - (H_\rho)^3 b(t)p(t) + 3(H_\rho)^2 H_{\rho\psi} b(t)p(t)\\
&+ 6(H_\rho)^2 H_{\rho\rho} b(t) - 3H_\rho H_{\rho\rho} b(t)p(t) + 3H_\rho b(t) H_{\rho\rho\psi} p(t) - (H_\psi)^2 r(t)p(t)\\
&+ H_\psi b(t) H_{\rho\rho\rho} p(t) + 3H_{\rho\psi} H_{\rho\rho} b(t)p(t) + H_\rho H_\psi p(t) + 4H_\rho b(t) H_{\rho\rho\rho} + 3(H_{\rho\rho})^2 b(t)\\
&- b(t)H_{\rho\rho\rho} p(t) + b(t) H_{\rho\rho\rho\psi} p(t) - r(t) H_{\psi\psi} p(t) + (H_\rho)^2 - H_\rho p(t) + H_{\rho\psi} p(t)\\
&+ b(t) H_{\rho\rho\rho\rho} + H_{\rho\rho} + 3H_\rho H_\psi H_{\rho\rho} b(t)p(t) = 0.
\end{aligned}$$

From assuming

$$H(\rho,\psi) = F(\xi), \qquad \xi = \rho,$$

which reduces to

$$r(t)e^{t+F(\xi)}\Big(F'(\xi)^2 + F''(\xi)\Big) = 0,$$

with the exact solution $F(\xi) = -t - \ln(-\frac{1}{b_1 + b_0(y-t)})$, we obtain

$$H(\rho,\psi) = -t - \ln(-\frac{1}{b_1 + b_0(y-t)}),$$

where b_0, b_1 are real constants. Thus, the exact solution of (1.81) can be evaluated as

$$u(x,y,t) = -b_1 - b_0(y-t).$$

Case (a.3)

For the case $r(t) = 0$ and $yh_1(t) + h_2(t) = 0$ we have

$$\left(u_t + a(t)(u^2)_x + b(t)u_{xxt}\right)_x = 0, \tag{1.83}$$

which satisfies the symmetry condition, namely

$$\xi = p(y), \qquad \eta = w(y), \qquad \tau = 0, \qquad \phi = uk(y).$$

Hence, the related infinitesimal generators are given by

$$V_1 = \frac{\partial}{\partial t}, \quad V_p = p(y)\frac{\partial}{\partial x}, \quad V_w = w(y)\frac{\partial}{\partial y}, \quad V_k = uk(y)\frac{\partial}{\partial u},$$

where the similarity variables can be extracted as follows:

$$\rho = x - \int g(y)dy, \qquad \psi = t - \int \frac{dy}{w(y)}, \qquad u = e^{\int s(y)dy + H(\rho,\psi)}, \tag{1.84}$$

where $g(y) = \frac{p(y)}{w(y)}$ and $s(y) = \frac{k(y)}{w(y)}$.

Substituting (1.84) in (1.76) yields

$$e^{\int s(y)dt + H(\rho,\psi)}\Big(H_\rho H_\psi + H_{\rho,\psi}\Big) + 2a(t)(H_\rho)^2 e^{2\int s(y)dy + 2H(\rho,\psi)}$$
$$+ 2a(t)e^{2\int s(y)dy + 2H(\rho,\psi)}\Big((H_\rho)^2 + H_{\rho\rho}\Big) + b(t)e^{\int s(y)dy + H(\rho,\psi)}\Big((H_\rho)^3 H_\psi$$
$$+ 3(H_\rho)^2 H_{\rho\psi} + 3H_\rho H_\psi H_{\rho\rho} + 3H_{\rho\rho\psi}H_\rho + H_{\rho\rho\rho}H_\psi + 3H_{\rho\rho}H_{\rho\psi} + H_{\rho\rho\rho\psi}\Big) = 0. \tag{1.85}$$

To reduce PDE (1.85), assume

$$H(\rho,\psi) = F(\xi), \qquad \xi = \rho,$$

which reduces (1.85) to

$$4a(t)F'(\xi)^2 e^{2\int s(y)dy + 2F(\xi)} - b(t)e^{\int s(y)dy + F(\xi)}F'(\xi)^3$$
$$+ 2a(t)e^{2\int s(y)dy + 2F(\xi)}F''(\xi) - 3b(t)e^{\int s(y)dy + F(\xi)}F''(\xi)F'(\xi)$$
$$- b(t)e^{\int s(y)dy + F(\xi)}F'''(\xi) - e^{\int s(y)dy + F(\xi)}F'(\xi) = 0. \tag{1.86}$$

The solution of (1.86) is $F(\xi) = -\frac{1}{2}\ln\left(-\frac{1}{b_1 + b_0(x - \int g(y)dy)}\right)$. So we have

$$H(\rho,\psi) = -\frac{1}{2}ln\Big(-\frac{1}{b_1 + b_0(x - \int g(y)dy)}\Big),$$

where b_0, b_1 is a real constant. Thus, the exact solution of (1.83) can be evaluated as

$$u(x,y,t) = e^{\int s(y)dy - \frac{1}{2}ln\left(-\frac{1}{b_1 + b_0(x - \int g(y)dy)}\right)}.$$

Case (b): $n = 3$. In this case, let us to consider the following equation

$$\left(u_t + a(t)(u^3)_x + b(t)u_{xxt}\right)_x + r(t)u_{yy} = 0. \tag{1.87}$$

The symmetry group of Eq.(1.87) will be generated by the vector field of the form (1.77). Applying the fourth prolongation $Pr^{(4)}$ to Eq.(1.87), we find

$$\begin{aligned} \xi &= b_1, & \eta &= a_3 + b_3 y + 2y^2, \\ \tau &= 0, & \phi &= u(b_2 + 2y + b_1) + x(ya_4 + b_4) + k(y,t), \end{aligned}$$

where $b_1, b_2, b_3, a_3, a_4, b_4$ are arbitrary constants. The infinitesimal generators can be considered by the following cases.

Case (b.1)

For $a(t) = 0$ we have

$$\left(u_t + b(t)u_{xxt}\right)_x + r(t)u_{yy} = 0, \tag{1.88}$$

with the symmetry condition

$$\begin{aligned} \xi &= b_1 + a_1 x, & \tau &= \frac{a_4}{r(t)} - \frac{\int_1^t a_1 r(\tau)d\tau}{r(t)} + \frac{2\int_1^t b_3 t(\tau)d\tau}{r(t)}, \\ \eta &= c_3 + b_3 y, & \phi &= u(d_1 + \frac{1}{2}b_3 + a_1) + g(x,y,t), \end{aligned}$$

where $a_1, a_4, b_1, b_3, c_3, d_1$ are arbitrary constants. So, the functions $b(t)$ and $r(t)$ are given by the following conditions:

$$b(t) = a_5\left(a_4 + \int_1^t -(a_1 - 2b_3)r(\tau)d\tau\right)^{\frac{-2a_1}{a_1 - 2b_3}}.$$

The infinitesimal generators are given by

$$\begin{aligned} V_1 &= u\frac{\partial}{\partial u} - \left(\frac{\int_1^t r(\tau)d\tau}{r(t)}\right)\frac{\partial}{\partial x} + x\frac{\partial}{\partial y}, \quad V_2 = \frac{\partial}{\partial y}, \\ V_3 &= \frac{u}{2}\frac{\partial}{\partial u} + 2\left(\frac{\int_1^t r(\tau)d\tau}{r(t)}\right)\frac{\partial}{\partial x} + y\frac{\partial}{\partial t}, \quad V_4 = \frac{1}{r(t)}\frac{\partial}{\partial x}, \\ V_5 &= u\frac{\partial}{\partial u}, \quad V_6 = \frac{\partial}{\partial t}, \quad V_7 = g(x,y,t)\frac{\partial}{\partial u}, \end{aligned}$$

where g is an arbitrary function satisfying Eq. (1.88). For the vector field

$$V_2 + V_4 + V_5 + V_6 = \frac{\partial}{\partial y} + \frac{1}{r(t)}\frac{\partial}{\partial x} + u\frac{\partial}{\partial u} + \frac{\partial}{\partial t},$$

we obtain

$$\rho = y - t, \qquad \psi = x - \int \frac{dt}{r(t)}, \qquad u = e^{t + H(\rho,\psi)},$$

from which similar to previous cases the exact solution of (1.88) can be evaluated as

$$u(x, y, t) = -b_1 - b_0(y - t),$$

where b_0, b_1 are real constants.

Case (b.2)

For $a(t) = r(t) = 0$ we conclude

$$\left(u_t + b(t)u_{xxt}\right)_x = 0. \tag{1.89}$$

A one-parameter Lie group of infinitesimal transformation of Eq. (1.89) is:

$$V_1 = u\frac{\partial}{\partial u} + \frac{\partial}{\partial x}, \quad V_2 = u\frac{\partial}{\partial u}, \quad V_3 = 2uy\frac{\partial}{\partial u}, \quad V_4 = xy\frac{\partial}{\partial u},$$
$$V_5 = x\frac{\partial}{\partial u}, \quad V_6 = \frac{\partial}{\partial x}, \quad V_7 = \frac{\partial}{\partial y}, \quad V_8 = y\frac{\partial}{\partial y},$$
$$V_9 = 2y^2\frac{\partial}{\partial y}, \quad V_k = k(y,t)\frac{\partial}{\partial u},$$

where $k(y, t)$ is an arbitrary function satisfying Eq. (1.89). For the vector field

$$V_5 + V_6 + V_7 = x\frac{\partial}{\partial u} + \frac{\partial}{\partial x} + \frac{\partial}{\partial y},$$

we have

$$\frac{dx}{1} = \frac{dy}{1} = \frac{dt}{0} = \frac{du}{x}, \tag{1.90}$$

and by solving Eq. (1.90) we obtain

$$\rho = x - y, \qquad \psi = t, \qquad u = \frac{1}{2}x^2 + H(\rho, \psi). \tag{1.91}$$

Substituting (1.91) into (1.89) yields

$$H_{\rho\psi} + b(t)H_{\rho\rho\rho\psi} = 0;$$

therefore,

$$H(\rho, \psi) = \mathcal{F}_1(\rho) + \mathcal{F}_2(\psi),$$

where $\mathcal{F}_1$ and $\mathcal{F}_2$ are arbitrary functions with respect to ρ and ψ, respectively. Thus, the exact solution of (1.89) can be evaluated as

$$u(x, y, t) = \frac{1}{2}x^2 + \mathcal{F}_1(x - y) + \mathcal{F}_2(t).$$

1.1.6 Lie symmetries of the mKdV-KP equation

In this section, we briefly discuss the symmetries of the mKdV-KP equation [172, 179]

$$\left(u_t - \frac{3}{2}u_x + 6u^2u_x + u_{xxx}\right)_x + u_{yy} = 0. \tag{1.92}$$

Let us consider a one-parameter Lie group of infinitesimal transformations:

$$\begin{cases} t^* = t + \epsilon\xi^1(t,x,y,u) + O(\epsilon^2), \\ x^* = x + \epsilon\xi^2(t,x,y,u) + O(\epsilon^2), \\ y^* = y + \epsilon\xi^3(t,x,y,u) + O(\epsilon^2), \\ u^* = u + \epsilon\phi(t,x,y,u) + O(\epsilon^2), \end{cases}$$

where ϵ is the group parameter. The associated Lie algebra of infinitesimal symmetries is the set of the vector field of the form

$$V = \xi^1(t,x,y,u)\frac{\partial}{\partial t} + \xi^2(t,x,y,u)\frac{\partial}{\partial x} + \xi^3(t,x,y,u)\frac{\partial}{\partial y} + \phi(t,x,y,u)\frac{\partial}{\partial u}.$$

If $Pr^{(4)}V$ denotes the fourth prolongation of V then the invariance condition

$$Pr^{(4)}V(\Delta)|_{\Delta=0} = 0,$$

where $\Delta := \left(u_t - \frac{3}{2}u_x + 6u^2u_x + u_{xxx}\right)_x + u_{yy}$ is equivalent to

$$\phi^{tx} + \left(6u^2 - \frac{3}{2}\right)\phi^{xx} + 12\left(u_x^2 + uu_{xx}\right)\phi + 24uu_x\phi^x + \phi^{xxxx} + \phi^{yy} = 0,$$

whenever $\Delta = 0$. Solving determining equations yields

$$\xi^1 = C_1 - 3tC_5,\ \xi^2 = C_2 + yC_4 + (3t - x)C_5,\ \xi^3 = C_3 - 2tC_4 - 2yC_5,\ \phi = uC_5,$$

where $C_i, (i = 1, \ldots, 5)$ are arbitrary constants. Thus, the Lie symmetry algebra admitted by Eq. (1.92) is spanned by the following five infinitesimal generators

$$V_1 = \frac{\partial}{\partial t},\ V_2 = \frac{\partial}{\partial x},\ V_3 = \frac{\partial}{\partial y},\ V_4 = y\frac{\partial}{\partial x} - 2t\frac{\partial}{\partial y},$$
$$V_5 = -3t\frac{\partial}{\partial t} + (3t - x)\frac{\partial}{\partial x} - 2y\frac{\partial}{\partial y} + u\frac{\partial}{\partial u}.$$

Also, corresponding generators of the optimal system of one-dimensional subalgebras are:
(1) $\mathcal{L}_{1,1} = \langle V_1 \rangle$,
(2) $\mathcal{L}_{2,1} = \langle \alpha V_1 + V_2 \rangle$,
(3) $\mathcal{L}_{3,1} = \langle \alpha V_1 + V_4 \rangle$,

(4) $\mathcal{L}_{4,1} = \langle \alpha V_1 + \beta V_2 + V_3 \rangle$,
(5) $\mathcal{L}_{5,1} = \langle \alpha V_1 + \beta V_2 + V_5 \rangle$,
where $\alpha, \beta \in \mathbb{R}$ are arbitrary.

In the following, we list the corresponding similarity variables and similarity solutions as well as the reduced PDEs obtained from the generators of the optimal system.

Reduction 1. Similarity variables related to $\mathcal{L}_{1,1}$ are $u(t,x,y) = F(x,y)$, where F satisfies the following equation:

$$2F_{yy} + 24FF_x^2 - 3F_{xx} + 12F^2F_{xx} + 2F_{xxxx} = 0. \tag{1.93}$$

Reduction 2. Using the subalgebra $\mathcal{L}_{2,1}$, we obtain the similarity variables and similarity solutions $u(t,x,y) = F(\xi,\eta),\ \ \xi = \frac{\alpha x - t}{\alpha},\ \eta = y$, and reduced PDE is:

$$24\alpha FF_\xi^2 - (2+3\alpha)F_{\xi\xi} + 12\alpha F^2F_{\xi\xi} + 2\alpha(F_{\eta\eta} + F_{\xi\xi\xi\xi}) = 0. \tag{1.94}$$

Reduction 3. Similarity variables and similarity solutions relating to $\mathcal{L}_{3,1}$ are $u(t,x,y) = F(\xi,\eta),\ \ \xi = \frac{\alpha y + t^2}{\alpha},\ \eta = \frac{3\alpha^2 x - 2t^3 - 3\alpha t y}{3\alpha^2}$, where

$$24\alpha FF_\eta^2 - (2\xi + 3\alpha)F_{\eta\eta} + 12\alpha F^2F_{\eta\eta} + 2\alpha(F_{\xi\xi} + F_{\eta\eta\eta\eta}) = 0. \tag{1.95}$$

Reduction 4. Using the subalgebra $\mathcal{L}_{4,1}$, we obtain the similarity variables and similarity solutions $u(t,x,y) = F(\xi,\eta),\ \ \xi = \frac{\alpha y - t}{\alpha},\ \eta = \frac{\alpha x - \beta t}{\alpha}$, and reduced PDE is:

$$24\alpha FF_\eta^2 - (2\beta + 3\alpha)F_{\eta\eta} + 12\alpha F^2F_{\eta\eta} + 2(\alpha F_{\xi\xi} - F_{\xi\eta} + \alpha F_{\eta\eta\eta\eta}) = 0. \tag{1.96}$$

Reduction 5. From the subalgebra $\mathcal{L}_{5,1}$, we have the similarity variables and similarity solutions $u(t,x,y) = \frac{F(\xi,\eta)}{\sqrt[3]{3t-\alpha}},\ \ \xi = \frac{y}{\sqrt[3]{(3t-\alpha)^2}},\ \eta = \frac{2x + 3t - 3\alpha - 2\beta}{2\sqrt[3]{3t-\alpha}}$, where

$$(\eta - 6F^2)F_{\eta\eta} + 2F_\eta + 2\xi F_{\xi\eta} - 12FF_\eta^2 - F_{\xi\xi} - F_{\eta\eta\eta\eta} = 0. \tag{1.97}$$

- **Solutions of reduced equations**

Here, we consider some reduced equations (1.93)-(1.97) of the previous section.

Eq. (1.93) admits two generators $\Gamma_1 = \frac{\partial}{\partial x}$ and $\Gamma_2 = \frac{\partial}{\partial y}$. Applying $\Gamma_1 + \Gamma_2$, the similarity variables are obtainable as follows:

$$F(x,y) = h(\xi),\ \ \xi = -x + y,$$

where

$$2h^{(4)} + (12h^2 - 1)h'' + 24h(h')^2 = 0. \tag{1.98}$$

Eq. (1.98) admits the only generator $\frac{\partial}{\partial \xi}$. Therefore it is not possible to solve it by current Lie symmetry methods. Here we want to apply the Nucci's reduction method to Eq. (1.98). This equation transforms into the following autonomous system of first order, i.e.,

$$\begin{cases} w_1' = w_2, \\ w_2' = w_3, \\ w_3' = w_4, \\ w_4' = \dfrac{(1-12w_1^2)w_3 - 24w_1w_2^2}{2}, \end{cases} \tag{1.99}$$

by the change of dependent variables

$$w_1(\xi) = h(\xi), \quad w_2(\xi) = h'(\xi), \quad w_3(\xi) = h''(\xi), \quad w_4(\xi) = h'''(\xi).$$

Let us choose w_1 as the new independent variable. Then (1.99) yields:

$$\begin{cases} \dfrac{dw_2}{dw_1} = \dfrac{w_3}{w_2}, \\ \dfrac{dw_3}{dw_1} = \dfrac{w_4}{w_2}, \\ \dfrac{dw_4}{dw_1} = \dfrac{(1-12w_1^2)w_3 - 24w_1w_2^2}{2w_2}. \end{cases} \tag{1.100}$$

From the first equation of (1.100) we have

$$w_3 = w_2\frac{dw_2}{dw_1},$$

which substituting into other equations of (1.100) yields:

$$\begin{cases} w_2\dfrac{d^2w_2}{dw_1^2} + \left(\dfrac{dw_2}{dw_1}\right)^2 = \dfrac{w_4}{w_2}, \\ (12w_1^2 - 1)\dfrac{dw_2}{dw_1} + 2\dfrac{dw_4}{dw_1} = 24w_1w_2. \end{cases} \tag{1.101}$$

From the first equation we conclude:

$$w_4 = w_2^2\frac{d^2w_2}{dw_1^2} + w_2\left(\frac{dw_2}{dw_1}\right)^2,$$

which by substituting in the second equation of (1.101) we have

$$2w_2^2\frac{d^3w_2}{dw_1^3} + 8w_2\frac{dw_2}{dw_1}\frac{d^2w_2}{dw_1^2} + 2\left(\frac{dw_2}{dw_1}\right)^3 + (12w_1^2 - 1)\frac{dw_2}{dw_1} + 24w_1w_2 = 0.$$

One time integration of obtained equation leads to

$$w_2^2\frac{d^2w_2}{dw_1^2} + 6w_1^2w_2 - \frac{1}{2}w_2 + w_2\left(\frac{dw_2}{dw_1}\right)^2 + c_1 = 0.$$

This equation doesn't admit any infinitesimal generators. It is possible to find the solution by supposition $c_1 = 0$. In this case we obtain:

$$w_2 = \frac{\sqrt{2w_1^2 - 4w_1^4 - 8c_2w_1 + 8c_3}}{2}. \tag{1.102}$$

In this step, we return the w_1 as a dependent variable, i.e., $w_1 = w_1(\xi)$. Therefore, by putting (1.102) in (1.99) we have a separable first order equation

$$w_1' = \frac{\sqrt{2w_1^2 - 4w_1^4 - 8c_2w_1 + 8c_3}}{2},$$

from which its implicit solution is easily obtainable. In a special case, by supposition $c_2 = 0$ an elliptic type of solution can be obtained as:

$$\frac{\sqrt{\left(16 - \frac{2\left(-1+\sqrt{32c_3+1}\right)w_1^2}{c_3}\right)\left(16 + \frac{2\left(1+\sqrt{32c_3+1}\right)w_1^2}{c_3}\right)}}{\sqrt{\frac{-1+\sqrt{32c_3+1}}{c_3}}\left(2w_1^2 - 4w_1^4 + 8c_3\right)} \times$$

$$EllipticF\left(\frac{\sqrt{2}w_1\sqrt{\frac{-1+\sqrt{32c_3+1}}{c_3}}}{4}, \frac{\sqrt{-16 - \frac{1+\sqrt{32c_3+1}}{c_3}}}{4}\right) = 2\sqrt{2}\xi + c_4.$$

Also, more restriction $c_3 = 0$ yields to an implicit form of solution:

$$\frac{w_1\sqrt{2 - 4w_1^2}}{\sqrt{2w_1^2 - 4w_1^4}} arctanh\left(\sqrt{\frac{1}{1 - 2w_1^2}}\right) = -\frac{1}{\sqrt{2}}\xi + c_4, \tag{1.103}$$

which setting $w_1 = u(t,x,y)$ and $\xi = -x + y$ in (1.103) gives a solution of Eq. (1.92) as:

$$\frac{u(t,x,y)\sqrt{2 - 4u(t,x,y)^2}}{\sqrt{2u(t,x,y)^2 - 4u(t,x,y)^4}} arctanh\left(\sqrt{\frac{2}{2 - 4u(t,x,y)^2}}\right) = \frac{x-y}{\sqrt{2}} + c_4.$$

Eq. (1.94) admits two infinitesimal generators $\Gamma_1 = \frac{\partial}{\partial \xi}$ and $\Gamma_2 = \frac{\partial}{\partial \eta}$. Also in the special case $\alpha = -\frac{2}{3}$, Eq. (1.94) admits another generator $\Gamma_3 = -2\xi\frac{\partial}{\partial \xi} - \eta\frac{\partial}{\partial \eta} + F\frac{\partial}{\partial F}$ in which here we just discuss the case $\alpha \in \mathbb{R}\backslash\{-\frac{2}{3}\}$. Invariance surface condition for $\Gamma_1 + \Gamma_2$ is $F_\xi + F_\eta = 0$; therefore similarity variables are

$$F(\xi,\eta) = h(\gamma), \quad \gamma = -\xi + \eta,$$

where $h(\gamma)$ satisfies the equation

$$2\alpha h^{(4)} + \left(12\alpha h^2 - \alpha - 2\right)h'' + 24\alpha h(h')^2 = 0. \tag{1.104}$$

Two times integration of (1.104) yields:

$$h'' + 2h^3 - \frac{(2+\alpha)h}{2\alpha} + c_1 + c_2\gamma = 0,$$

which by substituting $c_1 = c_2 = 0$ we obtain:

$$h(\gamma) = c_3\sqrt{\frac{2+\alpha}{2-\alpha+2c_3^2\alpha}}$$
$$\times\, JacobiSN\Big((\sqrt{\frac{\alpha-2}{2\alpha}}\gamma + c_4)\sqrt{\frac{2+\alpha}{2-\alpha+2c_3^2\alpha}}, c_3\frac{\sqrt{2\alpha(2-\alpha)}}{\alpha-2}\Big).$$

Back substituting of similarity variables yields:

$$u(t,x,y) = c_3\sqrt{\frac{2+\alpha}{2-\alpha+2c_3^2\alpha}}\times$$
$$JacobiSN\left((\sqrt{\frac{\alpha-2}{2\alpha}}\times\frac{t+\alpha(y-x)}{\alpha} + c_4)\sqrt{\frac{2+\alpha}{2-\alpha+2c_3^2\alpha}}, c_3\frac{\sqrt{2\alpha(2-\alpha)}}{\alpha-2}\right).$$

In a special case, by taking

$$c_3 = \frac{\alpha-2}{\sqrt{2\alpha(2-\alpha)}},$$

we obtain:

$$u(t,x,y) = -\frac{1}{2}\sqrt{\frac{2+\alpha}{\alpha}}\tanh\left(\frac{\sqrt{\frac{2\alpha+4}{2-\alpha}}\left(\sqrt{2\alpha(\alpha-2)}\times\frac{t+\alpha(y-x)}{\alpha} + 2c_4\alpha\right)}{4\alpha}\right).$$

The third reduced PDE, i.e., Eq. (1.95), admits the only symmetry $\Gamma = \frac{\partial}{\partial\eta}$. Using Γ we get the similarity variables and solutions of (1.95) as follows:

$$F(\xi,\eta) = h(\gamma), \quad \gamma = \xi,$$

and substituting $h(\gamma)$ in (1.95) we obtain:

$$h''(\gamma) = 0,$$

which two times integration leads to

$$h(\gamma) = c_1\gamma + c_2.$$

Thus, by back substituting the similarity variables we get

$$u(t,x,y) = c_1\left(\frac{\alpha y + t^2}{\alpha}\right) + c_2.$$

Also, the following special solutions are obtainable:

$$F(\xi,\eta) = c_4 \pm \sqrt{3}c_5 sech\,(c_3 + c_2\eta + c_1\xi) + c_5\tanh\,(c_3 + c_2\eta + c_1\xi)\,.$$

Back substituting the similarity variables gives the other solutions of Eq. (1.92) as follows:

$$u(t,x,y) = c_4 \pm \sqrt{3}c_5 sech\left(c_3 + c_2\frac{3\alpha^2 x - 2t^3 - 3\alpha t y}{3\alpha^2} + c_1\frac{\alpha y + t^2}{\alpha}\right)$$
$$+ c_5 \tanh\left(c_3 + c_2\frac{3\alpha^2 x - 2t^3 - 3\alpha t y}{3\alpha^2} + c_1\frac{\alpha y + t^2}{\alpha}\right).$$

Eq. (1.96) for $\alpha(1+6\alpha^2+4\alpha\beta) \neq 0$ admits two generators $\Gamma_1 = \frac{\partial}{\partial\eta}$ and $\Gamma_2 = 2\frac{\partial}{\partial\xi} - \frac{1}{\alpha}\frac{\partial}{\partial\eta}$. Invariance surface condition for $\Gamma_1+\Gamma_2$ is $2F_\xi + (1-\frac{1}{\alpha})F_\eta = 0$; therefore similarity variables are

$$F(\xi,\eta) = h(\gamma), \quad \gamma = \frac{(1-\alpha)\xi + 2\alpha\eta}{2\alpha},$$

where $h(\gamma)$ satisfies the following equation

$$4\alpha^2 h^{(4)} + \left(24\alpha^2 h^2 - 1 - 5\alpha^2 - 4\alpha\beta\right)h'' + 48\alpha^2 h(h')^2 = 0. \qquad (1.105)$$

Two times integration of (1.105) yields:

$$h'' + 2h^3 - \frac{(1+5\alpha^2+4\alpha\beta)h}{4\alpha^2} + c_1 + c_2\gamma = 0,$$

from which by substituting $c_1 = c_2 = 0$ we obtain:

$$h(\gamma) = c_3\sqrt{\frac{1+5\alpha^2+4\alpha\beta}{1+(1+4c_3^2)\alpha^2+4\alpha\beta}} \times$$
$$JacobiSN\left((\sqrt{\frac{-\alpha^2-4\alpha\beta-1}{2\alpha}}\gamma + c_4)\sqrt{\frac{1+5\alpha^2+4\alpha\beta}{1+(1+4c_3^2)\alpha^2+4\alpha\beta}}, \frac{2c_3\alpha}{\sqrt{\alpha^2+4\alpha\beta+1}}\right).$$

Back substituting of the similarity variables yields:

$$u(t,x,y) = c_3\sqrt{\frac{1+5\alpha^2+4\alpha\beta}{1+(1+4c_3^2)\alpha^2+4\alpha\beta}} \times$$
$$JacobiSN\left((\sqrt{\frac{-\alpha^2-4\alpha\beta-1}{2\alpha}}\Theta + c_4)\sqrt{\frac{1+5\alpha^2+4\alpha\beta}{1+(1+4c_3^2)\alpha^2+4\alpha\beta}}, \frac{2c_3\alpha}{\sqrt{\alpha^2+4\alpha\beta+1}}\right),$$

where

$$\Theta = x + \frac{\alpha(1-\alpha)}{2\alpha^2}y - \frac{1-\alpha+2\alpha\beta}{2\alpha^2}t.$$

In a special case, by taking

$$c_3 = \frac{\sqrt{\alpha^2+4\alpha\beta}}{2\alpha},$$

we obtain another solution of Eq. (1.92) as follows:

$$u(t,x,y) = \frac{\sqrt{2(5\alpha^2+4\alpha\beta+1)}}{4\alpha}\times$$
$$\tanh\left(\frac{\left(2\alpha c_4 + \left(x+\frac{\alpha(1-\alpha)}{2\alpha^2}y - \frac{1-\alpha+2\alpha\beta}{2\alpha^2}t\right)\sqrt{-1-\alpha-4\alpha\beta}\right)\sqrt{\frac{2(5\alpha^2+4\alpha\beta+1)}{\alpha^2+4\alpha\beta+1}}}{4\alpha}\right).$$

Finally, Eq. (1.97) does not admit the Lie symmetries and here we consider the translation in η, i.e.,

$$F(\xi,\eta) = h(\gamma), \quad \gamma = \xi,$$

where $h(\gamma)$ satisfies

$$h''(\gamma) = 0. \tag{1.106}$$

Therefore

$$h(\gamma) = c_1\gamma + c_2,$$

from which back substitution of variables yields the following solution

$$u(t,x,y) = \frac{c_1 y}{3t-\alpha} + \frac{c_2}{\sqrt[3]{3t-\alpha}}.$$

1.2 Nonclassical Lie symmetry analysis

In this section, we discuss a class of the point transformation groups which can lead to exact solutions of differential equations but are not symmetries. For some differential equations this method leads to new invariant solutions that cannot be extracted from the classical point symmetries. There are some different approaches to get the nonclassical Lie symmetries. Here, we introduce the nonclassical Lie symmetries by the *heir-equations* method.

1.2.1 Nonclassical symmetries for a class of reaction-diffusion equations

In [25, 88, 38] a class of reaction-diffusion equations, i.e.,

$$u_t = u_{xx} + cu_x + R(u,x), \tag{1.107}$$

with $R(u,x)$ arbitrary function of u and x, was introduced as a model that incorporates climate shift, population dynamics and migration for a population of individuals $u(t,x)$ that reproduce, disperse and die within a patch of favorable habitat surrounded by unfavorable habitat. It is assumed that due to a shifting climate, the patch moves with a fixed speed $c > 0$ in a one-dimensional universe.[4]

Motivated by this study here we look for nonclassical symmetries of equation (1.107) with the purpose of finding explicit expressions of the function $R(u, x)$ and deriving nonclassical symmetry solutions when feasible.

Nonclassical symmetries were introduced in 1969 in a seminal paper by Bluman and Cole [14].

One should be aware that some authors call nonclassical symmetries Q-conditional symmetries[5] of the second type, e.g., [37], while others call them reduction operators, e.g., [157].

The nonclassical symmetry method can be viewed as a particular instance of the more general differential constraint method that, as stated by Kruglikov [115], *dates back at least to the time of Lagrange... and was introduced into practice by Yanenko* [184]. The method was set forth in details in Yanenko's monograph [170] that was not published until after his death [48]. A more recent account and generalization of Yanenko's work can be found in [131].

Among the papers dedicated to the application of the nonclassical symmetry method to diffusion-convection equations with source, we single out [32] where some nonclassical symmetries solutions were determined for the equation:

$$u_t = u_{xx} + k(x)u^2(1 - u). \tag{1.108}$$

In particular nonclassical symmetries of the type $V(t, x)\frac{\partial}{\partial_x} + \frac{\partial}{\partial_t}$ were found in the following three instances:

$$(i)\ \ k(x) = a^2x^2, \qquad (ii)\ \ k(x) = a^2 \tanh^2 x, \qquad (iii)\ \ k(x) = a^2 \tan^2 x, \tag{1.109}$$

with a arbitrary constant.

In the next section, we introduce the concept of heir-equations [141] and their link to nonclassical symmetries [144].

1.2.1.1 Heir-equations and nonclassical symmetries

Let us consider an evolution equation in two independent variables and one dependent variable of second order:

$$u_t = H(t, x, u, u_x, u_{xx}). \tag{1.110}$$

[4]Actually equation (1.107) corresponds to the original model

$$u_t = u_{zz} + R(u, z - ct)$$

rewritten in terms of a moving coordinate system with $x = z - ct$ [25].

[5]In [59] this name was introduced for the first time.

If

$$\Gamma = V_1(t,x,u)\frac{\partial}{\partial t} + V_2(t,x,u)\frac{\partial}{\partial x} - F(t,x,u)\frac{\partial}{\partial u} \tag{1.111}$$

is a generator of a Lie point symmetry[6] of equation (1.110) then the invariant surface condition is given by:

$$V_1(t,x,u)u_t + V_2(t,x,u)u_x = F(t,x,u). \tag{1.112}$$

Let us take the case with $V_1 = 0$ and $V_2 = 1$, so that (1.112) becomes:[7]

$$u_x = G(t,x,u). \tag{1.113}$$

Then, an equation for G is easily obtained. We call this equation G-equation [140]. Its invariant surface condition is given by:

$$\xi_1(t,x,u,G)G_t + \xi_2(t,x,u,G)G_x + \xi_3(t,x,u,G)G_u = \eta(t,x,u,G). \tag{1.114}$$

Let us consider the case $\xi_1 = 0$, $\xi_2 = 1$, and $\xi_3 = G$, so that (1.114) becomes:

$$G_x + GG_u = \eta(t,x,u,G).$$

Then, an equation for η is derived. We call this equation η-equation. Clearly:

$$G_x + GG_u \equiv u_{xx} \equiv \eta.$$

We could keep iterating to obtain the Ω-equation, which corresponds to:

$$\eta_x + G\eta_u + \eta\eta_G \equiv u_{xxx} \equiv \Omega(t,x,u,G,\eta),$$

the ρ-equation, which corresponds to:

$$\Omega_x + G\Omega_u + \eta\Omega_G + \Omega\Omega_\eta \equiv u_{xxxx} \equiv \rho(t,x,u,G,\eta,\Omega),$$

and so on. Each of these equations inherits the symmetry algebra of the original equation, with the right prolongation: first prolongation for the G-equation, second prolongation for the η-equation, and so on. Therefore, these equations were named heir-equations in [141]. This implies that even in the case of few Lie point symmetries many more Lie symmetry reductions can be performed by using the invariant symmetry solution of any of the possible heir-equations, as it was shown in [141, 13, 129].

We recall that the heir-equations are just some of the many possible n-extended equations as defined by Guthrie in [69].

In [68] Goard has shown that Nucci's method of constructing heir-equations by iterating the nonclassical symmetries method is equivalent to the generalized conditional symmetries method.

[6] The minus sign in front of $F(t,x,u)$ was put there for the sake of simplicity; it could be replaced with a plus sign without affecting the following results.

[7] We have replaced $F(t,x,u)$ with $G(t,x,u)$ in order to avoid any ambiguity in the following discussion.

The difficulty in applying the method of nonclassical symmetries consists in solving nonlinear determining equations in contrast with the linearity of the determining equations in the case of classical symmetries.

The concept of Gröbner basis has been used [41] for this purpose.

In [144] it was shown that one can find the nonclassical symmetries of any evolution equations of any order by using a suitable heir-equation and searching for a given particular solution among all its solutions, thus avoiding any complicated calculations. We recall the method as applicable to equation (1.110).

We derive u_t from (1.110) and replace it into (1.112), with the condition $V_1 = 1$, i.e.,

$$H(t, x, u, u_x, u_{xx}) + V_2(t, x, u)u_x = F(t, x, u). \tag{1.115}$$

Then, we generate the η-equation with $\eta = \eta(x, t, u, G)$, and replace $u_x = G$, $u_{xx} = \eta$ into (1.115), i.e.,

$$H(t, x, u, G, \eta) = F(t, x, u) - V_2(t, x, u)G. \tag{1.116}$$

For Dini's theorem, we can isolate η in (1.116), e.g.,

$$\eta = [h_1(t, x, u, G) + F(t, x, u) - V_2(t, x, u)G]\, h_2(t, x, u, G), \tag{1.117}$$

where $h_i(t, x, u, G)(i = 1, 2)$ are known functions. Thus, we have obtained a particular solution of η which must yield an identity if replaced into the η-equation. The only unknowns are $V_2 = V_2(t, x, u)$ and $F = F(t, x, u)$. If any such solution is singular, i.e., does not form a group then we have found the nonclassical symmetries; otherwise one obtains the classical symmetries [144].

We use a simple MAPLE program to derive the heir-equations. In particular the G-equation of (1.107) is

$$G_t + RG_u - G_{xx} - 2GG_{xu} - G^2G_{uu} - cG_x - R_uG - R_x = 0,$$

and the η-equation is

$$\begin{aligned}\eta_t + R\eta_u + R_uG\eta_G - \eta_{xx} - 2G\eta_{xu} - 2\eta\eta_{xG} - G^2\eta_{uu} - 2G\eta\eta_{uG}\\ - \eta^2\eta_{GG} - c\eta_x - R_{uu}G^2 - R_u\eta - 2GR_{xu} + R_x\eta_G - R_{xx} = 0.\end{aligned} \tag{1.118}$$

The particular solution of the η-equation that we are looking for is

$$\eta(t, x, u, G) = -R(u, x) - cG + F(t, x, u) - V_2(t, x, u)G, \tag{1.119}$$

that replaced into (1.140) yields an overdetermined system in the unknowns F, V_2 and $R(u, x)$. Since we obtain a polynomial of third degree in G then we let MAPLE evaluate the four coefficients that we call d_i, $i = 0, 1, 2, 3$ where i stands for the corresponding power of u. We impose all of them to be zero. From d_3, we obtain

$$V_2(t, x, u) = ss_1(t, x)u + ss_2(t, x),$$

while d_2 yields

$$F(t,x,u) = -\frac{1}{3}ss_1^2u^3 + \frac{1}{2}\left(\frac{\partial ss_1}{\partial x} - 2css_1 - 2ss_1ss_2\right) + ss_3(t,x)u + ss_4(t,x),$$

with $ss_j(t,x),\ \ j = 1,\ldots,4$ arbitrary functions of t and x. Then after differentiating d_1 four times with respect to u we obtain

$$\frac{\partial^4 R(u,x)}{\partial u^4} = 0, \tag{1.120}$$

which implies that $R(u,x)$ must be a polynomial in u of third degree at most, i.e.,

$$R(u,x) = -\frac{a_3^2(x)}{6}u^3 + \frac{a_2(x)}{2}u^2 + a_1(x)u + a_0(x),$$

where $a_i(x),\ \ i = 0,1,2,3$ are arbitrary functions of x. Since none of the remaining arbitrary functions depends on u, and d_1 has now become a polynomial of degree 3 in u, we have to annihilate all the four coefficients, i.e., $d_{1,i},\ \ i = 0,1,2,3$. From $d_{1,3}$ we have that $ss_1(t,x)$ must be a constant, and two cases arise:

Case 1. $ss_1 = \pm\frac{\sqrt{3}}{2}\,a_3(x)$,

Case 2. $ss_1 = 0$.

We discuss the two cases,[8] separately. We remark that $a_3(x) = 0$ corresponds to a subcase of Case 2, and consequently in Case 1 we assume $a_3(x) \neq 0$.

Interestingly enough in Case 2 nonclassical symmetries exist for

$$R(u,x) = \frac{f(u)}{k^2(x)}$$

with $f(u)$ any arbitrary function of u, and $k(x)$ either of the following three particular functions of x, i.e.,

$$k(x) = -\frac{cx+2}{2x}, \quad k(x) = \frac{c}{e^{c(b_0-x)}-1}, \quad k(x) = \frac{1}{b_1}\tan\left(\frac{x+b_2}{b_1}\right) - \frac{c}{2}. \tag{1.121}$$

Case 1. $R(u,x) = -\frac{a_3^2(x)}{6}u^3 + \frac{a_2(x)}{2}u^2 + a_1(x)u + a_0(x)$.

From coefficients $d_{1,2}$, $d_{1,1}$, $d_{1,0}$ we obtain ss_2, ss_3 and ss_4, respectively. All of them are functions of x only, e.g.,

$$ss_2 = -\frac{1}{2a_3(x)}\left(-4a_3'(x) + \sqrt{3}a_2(x) + 2ca_3(x)\right),$$

[8] In Case 1, one can choose either the plus or minus sign indifferently.

where $'$ denotes differentiation with respect to x. Now the only remaining coefficient is d_0 which has become a linear polynomial in u. Therefore, we are left with two expressions to annihilate, namely the following undetermined system of two equations that contain the derivative of $a_3(x)$ up to fifth order, and fourth order, respectively, and lower derivatives of the other three functions $a_2(x)$, $a_1(x)$ and $a_0(x)$

$$
\begin{aligned}
&-a_3^3a_3^{(iv)}-4a_3^3a_3''a_1-c^2a_3^3a_3''-5a_2^2a_3a_3''+3a_2a_3^2a_2''+a_3^3\sqrt{3}a_2'''-36a_3(a_3')^2a_3''\\
&-2a_3^3a_3'''c+8a_3^2a_3'a_3'''+a_0'\sqrt{3}a_3^5-2a_3'a_3^3a_1'-18a_3'^3\sqrt{3}a_2+ca_3^4a_1'+4a_3^2a_3'^2a_1\\
&-a_2^3\sqrt{3}a_3'-2a_2\sqrt{3}a_3^2a_3''c+16a_3a_3'a_3''\sqrt{3}a_2+14a_3^2a_3'a_3''c+3a_3^2a_2'^2+5a_3^2a_3''^2\\
&+ca_3^3\sqrt{3}a_2''-5a_3^2a_3'\sqrt{3}a_2''-6a_3^2a_3''\sqrt{3}a_2'-2a_2\sqrt{3}a_3^2a_3'''+13a_3'^2a_2^2+a_1'a_3^3\sqrt{3}a_2\\
&+a_3^3\sqrt{3}a_2'a_1+14a_3a_3'^2\sqrt{3}a_2'-a_2^2a_3ca_3'+a_0\sqrt{3}a_3^4a_3'+a_2^2\sqrt{3}a_3a_2'\\
&-12a_3a_3'^3c+c^2a_3^2a_3'^2+4a_3a_3'^2c\sqrt{3}a_2-a_3^2a_3'\sqrt{3}a_2a_1-3a_3^2ca_3'\sqrt{3}a_2'\\
&+24a_3'^4-14a_3a_3'a_2a_2'+a_2a_3^2ca_2'+3a_3^4a_1''=0,
\end{aligned}
\tag{1.122}
$$

$$
\begin{aligned}
&\frac{6}{\sqrt{3}}a_3^4a_3^{(v)}+7c\sqrt{3}a_3^2a_3'a_2'a_2-ca_3^6a_0'-3a_3'^2a_2^3+192\sqrt{3}a_3'^5-26c\sqrt{3}a_3^3a_3'''a_3'\\
&+2c\sqrt{3}a_3^4a_1'a_3'-264a_3'^4a_2+5a_3^3a_3^{(iv)}a_2+104\sqrt{3}a_3'^2a_3^2a_3'''-18\sqrt{3}a_3^3a_3^{(iv)}a_3'\\
&+2c^2\sqrt{3}a_3^4a_3'''+4c\sqrt{3}a_3^4a_3^{(iv)}-2c\sqrt{3}a_3^5a_1''-26c\sqrt{3}a_3^3a_3''^2+19ca_3^3a_2''a_3'\\
&+27ca_3^3a_2'a_3''+6ca_3^3a_3'''a_2+c^2a_3^3a_3''a_2-416\sqrt{3}a_3'^3a_3a_3''+10\sqrt{3}a_3^4a_3''a_1'\\
&+178\sqrt{3}a_3^2a_3''^2a_3'-42\sqrt{3}a_3^3a_3''a_3'''-6a_3^3\sqrt{3}a_2'a_2''+\sqrt{3}a_3^2a_3'''a_2^2-\sqrt{3}a_3^3a_2'''a_2\\
&+350a_3'^2a_3a_3''a_2-204a_3^2a_3''a_2'a_3'-56a_3'a_2a_3^2a_3'''+2a_1\sqrt{3}a_3^4a_3'''+5a_1a_3^3a_3''a_2\\
&-54ca_3^2a_3'a_3''a_2-c^2a_3^4a_2''-4ca_3^4a_2'''-2\sqrt{3}a_3^5a_1'''+3a_0a_3^5a_3''-88a_3'^2a_3^2a_2''\\
&+39a_3^3a_3''a_2''-55a_3^2a_3''^2a_2+24a_3^3a_2'a_3'''+20a_3^3a_2'''a_3'-3a_1a_3^4a_2''-4a_0'a_3^5a_3'\\
&+34\sqrt{3}a_3'^3a_2^2-4a_3'^2a_3^4a_0+232a_3'^3a_3a_2'-6a_3^4a_2'a_1'-3a_3^2a_2'^2a_2-a_1a_3^4ca_2'\\
&+6c^2\sqrt{3}a_3^2a_3'^3-2c\sqrt{3}a_3^3a_2'^2-64c\sqrt{3}a_3a_3'^4+3c^2a_3^3a_3'a_2'-3c^2a_3^2a_3'^2a_2\\
&-56ca_3^2a_3'^2a_2'+62ca_3a_3'^3a_2-12\sqrt{3}a_3'^2a_3^3a_1'+16a_3^2\sqrt{3}a_2'^2a_3'-a_0a_3^5a_3'c\\
&+a_0'a_3^5\sqrt{3}a_2+6a_3a_2'a_3'a_2^2+6a_3'a_2a_3^3a_1'+a_1a_3^3ca_3'a_2-50a_3\sqrt{3}a_2'a_3'^2a_2\\
&-5c\sqrt{3}a_3a_3'^2a_2^2+a_3'\sqrt{3}a_2a_3^4a_0-2a_1\sqrt{3}a_3^5a_1'+8a_1\sqrt{3}a_3^2a_3'^3+8a_1a_3^3a_2'a_3'\\
&-10a_1a_3^2a_3'^2a_2-a_1\sqrt{3}a_3^3a_2'a_2-2a_1\sqrt{3}a_3^3ca_3'^2+a_1\sqrt{3}a_3^2a_3'a_2^2-3a_3^6a_0''\\
&-3a_3^4a_2^{(iv)}-c\sqrt{3}a_3^3a_2''a_2+11a_3'\sqrt{3}a_2a_3^2a_2''+16\sqrt{3}a_3^2a_3''a_2'a_2-8c^2\sqrt{3}a_3^3a_3'a_3''\\
&+112c\sqrt{3}a_3^2a_3'^2a_3''+c\sqrt{3}a_3^2a_3''a_2^2-21\sqrt{3}a_3a_3''a_3'a_2^2-10a_1\sqrt{3}a_3^3a_3''a_3'\\
&+2a_1\sqrt{3}a_3^4ca_3''+4\sqrt{3}a_3^4a_1''a_3'=0.
\end{aligned}
\tag{1.123}
$$

Since this system has infinite solutions we look for some particular ones.

1.2.1.2 $R(u,x) = -\frac{1}{2}x^2u^3 + 3u^2 + \frac{1}{2}c^2u$

If we assume $a_3(x) = \sqrt{3}x$, and $a_2(x)$, $a_1(x), a_0(x)$ to be constants then from system (1.122)-(1.123) we obtain that

$$R(u,x) = -\frac{1}{2}x^2u^3 + 3u^2 + \frac{1}{2}c^2u, \tag{1.124}$$

and

$$ss_1(t,x) = \frac{3x}{2}, \quad ss_2(t,x) = -\frac{1+cx}{x}, \quad ss_3(t,x) = c\frac{-2+3cx}{4x}, \quad ss_4(t,x) = 0.$$

Thus, (1.119) becomes

$$\eta = -\frac{x^3u^3 + 2cu - c^2xu + 6x^2uG - 4G}{4x},$$

namely

$$u_{xx} = -\frac{x^3u^3 + 2cu - c^2xu + 6x^2uu_x - 4u_x}{4x},$$

that can be solved in closed form, i.e.,

$$u(t,x) = \frac{c^2R_2(t)e^{\frac{cx}{2}} - c^2(1+cx)e^{\frac{-cx}{2}}}{R_1(t) + (cx-2)R_2(t)e^{\frac{cx}{2}} + (10+5cx+c^2x^2)e^{\frac{-cx}{2}}}, \tag{1.125}$$

with $R_k(t), k = 1,2$ arbitrary functions of t. Substituting (1.125) into (1.107) yields the following nonclassical symmetry solution

$$u(t,x) = \frac{c^2c_1e^{c^2t+\frac{cx}{2}} - c^2(1+cx)e^{\frac{-cx}{2}}}{c_2e^{\frac{-c^2t}{4}} + c_1(cx-2)e^{c^2t+\frac{cx}{2}} + (10+5cx+c^2x^2)e^{\frac{-cx}{2}}}, \tag{1.126}$$

with $c_k, k = 1,2$ arbitrary constants. We observe that

$$\lim_{t\to\infty} u(t,x) = \frac{c^2}{cx-2}, \quad \lim_{x\to\pm\infty} u(t,x) = 0,$$

and that $u(t,x) < 0$ for $t > 0, x < 0$. This means that the solution (1.126) is not defined at $x = 2/c$ and is positive[9] if $x \geq 0$.

1.2.1.3 $R(u,x) = -\frac{1}{2}e^{cx}u^3 + \frac{c^2}{4}u + e^{\frac{cx}{2}}$

If we assume $a_3(x) = \sqrt{3}e^{cx}$, $a_2(x) = 0$, and $a_1(x) = b_1, a_0(x) = b_0$, i.e., constants, then from system (1.122)-(1.123) we obtain that

$$R(u,x) = -\frac{1}{2}e^{cx}u^3 + b_1u + b_0e^{\frac{cx}{2}}, \quad [b_1, b_0 = \text{const.}] \tag{1.127}$$

[9]It depends also on the values given to the arbitrary constants.

and thus η becomes

$$\eta = -\frac{1}{8}\left(2e^{cx}u^3 + (3c^2 - 4b_1)u + 8cG + 6e^{\frac{cx}{2}}cu^2 + 12e^{\frac{cx}{2}}uG - 4b_0e^{-\frac{cx}{2}}\right),$$

namely

$$u_{xx} = -\frac{1}{8}\left(2e^{cx}u^3 + (3c^2 - 4b_1)u + 8cu_x + 6e^{\frac{cx}{2}}cu^2 + 12e^{\frac{cx}{2}}uu_x - 4b_0e^{-\frac{cx}{2}}\right), \tag{1.128}$$

that can be solved in closed form, although the solution is very lengthy. If we assume

$$b_1 = \frac{c^2}{4}, \quad b_0 = 1,$$

then the solution of (1.128) becomes:

$$u(t,x) = \frac{\sqrt[3]{2}}{2\left(R_1(t)e^{\frac{\sqrt[3]{2}x}{2}} - R_2(t)e^{-\frac{\sqrt[3]{2}x}{4}}\sin\left(\frac{\sqrt[3]{2}\sqrt{3}x}{4}\right) + e^{-\frac{\sqrt[3]{2}x}{4}}\cos\left(\frac{\sqrt[3]{2}\sqrt{3}x}{4}\right)\right)e^{\frac{cx}{2}}}$$
$$\times\left[2R_1(t)e^{\frac{\sqrt[3]{2}x}{2}} + R_2(t)e^{\frac{-\sqrt[3]{2}x}{4}}\left(\sin\left(\frac{\sqrt[3]{2}\sqrt{3}x}{4}\right) - \sqrt{3}\cos\left(\frac{\sqrt[3]{2}\sqrt{3}x}{4}\right)\right)\right.$$
$$\left.-e^{\frac{-\sqrt[3]{2}x}{4}}\left(\sqrt{3}\sin\left(\frac{\sqrt[3]{2}\sqrt{3}x}{4}\right) + \cos\left(\frac{\sqrt[3]{2}\sqrt{3}x}{4}\right)\right)\right],$$

which if replaced into (1.107) yields

$$R_1(t) = 0, \quad R_2(t) = -\tan\left(\frac{3\sqrt{3}\sqrt[3]{4}}{8}(t + c_1)\right).$$

This solution oscillates between negative and positive values. Consequently, it is not a valid solution for the biological model set in [25]. However, equation

$$u_t = u_{xx} + cu_x - \frac{1}{2}e^{cx}u^3 + \frac{c^2}{4}u + e^{\frac{cx}{2}}$$

may be of interest for other biological or physical models.

As the special case for $Case1$, we suppose

$$R(u,x) = -\frac{u^3}{6} - \frac{\sqrt{3}u^2}{x} + \frac{c^2u}{6} + \frac{\sqrt{3}c^2}{3x}.$$

Then, if we assume $a_3(x) = b_3, a_1(x) = b_1$ and substitute them into system (1.122)-(1.123), after some further simplifications such as $b_1 = c^2/6$ and having to impose that[10] $b_3 = -1$, then we obtain

$$R(u,x) = -\frac{u^3}{6} - \frac{\sqrt{3}u^2}{x} + \frac{c^2u}{6} + \frac{\sqrt{3}c^2}{3x}.$$

[10] It is also possible to have $b_3 = 1$ although it leads to very lengthy calculations.

Thus, (1.119) becomes

$$\eta = \frac{36xG - 36u + 6\sqrt{3}Gux^2 - u^3x^2 - 6\sqrt{3}u^2x + c^2ux^2 + 12\sqrt{3}c + 2\sqrt{3}c^2x}{12x^2},$$

namely

$$u_{xx} = \frac{36xu_x - 36u + 6\sqrt{3}u_xux^2 - u^3x^2 - 6\sqrt{3}u^2x + c^2ux^2 + 12\sqrt{3}c + 2\sqrt{3}c^2x}{12x^2}.$$

If we assume $c = 0$ then we find that its solution is

$$u(t,x) = -\frac{2\sqrt{3}(4R_2(t)x^3 + 2x)}{R_1(t) + R_2(t)x^4 + x^2},$$

which substituted into (1.107) yields the following solution

$$u(t,x) = \frac{4\sqrt{3}x(2x^2 + c_1 + 12t)}{6c_1t + 36t^2 - c_2 - x^4 - x^2c_1 - 12tx^2}. \tag{1.129}$$

Although this solution is not valid for the biological problem set in [25] since $c = 0$, we are reporting it since equation

$$u_t = u_{xx} + \frac{u^2(xu + 6\sqrt{3})}{6x}$$

may be of interest for other biological or physical problems. We observe that solution (1.129) obeys

$$\lim_{t\to\infty} u(t,x) = 0, \quad \lim_{x\to\pm\infty} u(t,x) = 0.$$

Moreover (1.129) is not defined for the following set of values of x and t:

$$\left\{x = \frac{1}{2}\sqrt{-2c_1 - 24t + 2\sqrt{c1^2 + 48c_1t + 288t^2 - 4c_2}}, \forall t\right\},$$

$$\left\{x = -\frac{1}{2}\sqrt{-2c_1 - 24t + 2\sqrt{c1^2 + 48c_1t + 288t^2 - 4c_2}}, \forall t\right\},$$

$$\left\{x = \frac{1}{2}\sqrt{-2c_1 - 24t - 2\sqrt{c1^2 + 48c_1t + 288t^2 - 4c_2}}, \forall t\right\},$$

$$\left\{x = -\frac{1}{2}\sqrt{-2c_1 - 24t - 2\sqrt{c1^2 + 48c_1t + 288t^2 - 4c_2}}, \forall t\right\}.$$

Also, if we suppose

$$R(u,x) = -u^3 + 6\frac{u}{x^2} + 6\frac{\sqrt{2}}{x^3}, \ [c = 0]$$

then we impose $a_3(x) = b_3, a_2(x) = 0$, and $c = 0$, so we obtain

$$R(u,x) = -u^3 + 6\frac{u}{x^2} + 6\frac{\sqrt{2}}{x^3}. \tag{1.130}$$

Thus (1.119) becomes

$$\eta = -\frac{6\sqrt{2} - 6xu + 3\sqrt{2}x^3uG + x^3u^3}{2x^3},$$

namely

$$u_{xx} = -\frac{6\sqrt{2} - 6xu + 3\sqrt{2}x^3uu_x + x^3u^3}{2x^3}.$$

We find its solution, i.e.,

$$u(t,x) = \frac{\sqrt{2}(-R_1(t) + 3R_2(t)x^4 + x^2)}{x(R_1(t) + R_2(t)x^4 + x^2)},$$

which substituted into (1.107) yields the following solution

$$u(t,x) = -\frac{3\sqrt{2}\left(12c_2^2 + 24c_2t + 4c_2x^2 + 4c_1 + 12t^2 + 4tx^2 + x^4\right)}{x\left(36c_2^2 + 72c_2t - 12c_2x^2 + 12c_1 + 36t^2 - 12tx^2 - x^4\right)}. \quad (1.131)$$

Although this solution is not valid for the biological problem set in [25] since $c = 0$, we report it here because the equation

$$u_t = u_{xx} - u^3 + 6\frac{u}{x^2} + 6\frac{\sqrt{2}}{x^3}, \quad (1.132)$$

may prove to be of interest for other biological or physical complex problems. We observe that solution (1.131) is such that

$$\lim_{t\to\infty} u(t,x) = -\frac{\sqrt{2}}{x}, \qquad \lim_{x\to\pm\infty} u(t,x) = 0,$$

and that $u(t,x) < 0$ for $t > 0, x < 0$. Moreover (1.131) is not defined for the following set of values of x and t:

$$x = 0,$$

$$\left\{x = \sqrt{-6t - 6c_2 + 2\sqrt{18t^2 + 36tc_2 + 18c_2^2 + 3c_1}}, \forall t\right\},$$

$$\left\{x = -\sqrt{-6t - 6c_2 + 2\sqrt{18t^2 + 36tc_2 + 18c_2^2 + 3c_1}}, \forall t\right\},$$

$$\left\{x = \sqrt{-6t - 6c_2 - 2\sqrt{18t^2 + 36tc_2 + 18c_2^2 + 3c_1}}, \forall t\right\},$$

$$\left\{x = -\sqrt{-6t - 6c_2 - 2\sqrt{18t^2 + 36tc_2 + 18c_2^2 + 3c_1}}, \forall t\right\},$$

$$\left\{\forall x,\ t = \frac{1}{6}x^2 - c_2 + \frac{1}{6}\sqrt{2x^4 - 12c_1}\right\},$$

$$\left\{\forall x,\ t = \frac{1}{6}x^2 - c_2 - \frac{1}{6}\sqrt{2x^4 - 12c_1}\right\}.$$

Case 2. $R(u,x) = \dfrac{f(u)}{k^2(x)}$.

If we assume $ss_1 = 0$ then $V_2(t,x,u) = ss_2(t,x)$ and d_0 yields that

$$R(u,x) = \frac{f(u)}{k^2(x)}, \quad ss_2(t,x) = k(x).$$

Following [32] we impose $F(t,x,u) = 0$; thus the annihilation of d_1 imposes that $k(x)$ is either one of the three particular functions of x in (1.121) and their nonclassical symmetry operators are

$$\frac{\partial}{\partial t} - \frac{cx+2}{2x}\frac{\partial}{\partial x}, \; \frac{\partial}{\partial t} + \frac{c}{e^{c(b_0-x)}-1}\frac{\partial}{\partial x}, \; \frac{\partial}{\partial t} + \left(\frac{1}{b_1}\tan\left(\frac{x+b_2}{b_1}\right) - \frac{c}{2}\right)\frac{\partial}{\partial x},$$

respectively. In each case we can solve the corresponding invariant surface condition (1.112) and reduce the original diffusion equation (1.107) to an ordinary differential equation that involves an arbitrary function of the unknown due to the arbitrariness of $f(u)$. We consider some instances where $f(u)$ has a given expression in order to derive the nonclassical symmetry solution of equation (1.107).

For the Case 2, we assume $k(x) = -\dfrac{cx+2}{2x}$. We solve the invariant surface equation (1.112), i.e.,

$$u_t - \frac{cx+2}{2x}u_x = 0,$$

and derive its complete solution as

$$u(t,x) = H(\xi), \quad \xi = \frac{4\log(cx+2) - 2cx - c^2 t}{c^2},$$

where $H(\xi)$ is an arbitrary function of ξ. After substituting this solution into the equation (1.107), i.e.,

$$u_t = u_{xx} + cu_x + \frac{4x^2}{(cx+2)^2}f(u), \tag{1.133}$$

we obtain the following ordinary differential equation

$$4\frac{\mathrm{d}^2 H}{\mathrm{d}\xi^2} - c^2\frac{\mathrm{d}H}{\mathrm{d}\xi} + 4f(H) = 0. \tag{1.134}$$

Let us consider $f(u) = 1/u$. In this instance the equation (1.134) admits a two-dimensional non-Abelian transitive Lie point symmetry algebra[11] generated by

$$\frac{\partial}{\partial \xi}, \quad e^{\frac{c^2\xi}{4}}\left(4\frac{\partial}{\partial \xi} + c^2 H\frac{\partial}{\partial H}\right),$$

[11]We recall that the classification of real two-dimensional Lie symmetry algebra and derivation of corresponding canonical variables were done by Lie himself [119], retold in Bianchi's 1918 textbook [26] and also in more recent textbooks, e.g., [94].

and equation (1.134) can be integrated by quadrature. In fact taking a canonical representation of the generators of the two-dimensional Lie point symmetry algebra, i.e.,

$$4e^{\frac{c^2\xi}{4}}\left(4\frac{\partial}{\partial\xi}+c^2H\frac{\partial}{\partial H}\right),\quad -\frac{4}{c^2}\frac{\partial}{\partial\xi}+4e^{\frac{c^2\xi}{4}}\left(4\frac{\partial}{\partial\xi}+c^2H\frac{\partial}{\partial H}\right),$$

we can derive the corresponding canonical variables, i.e.,

$$\tilde{\xi}=He^{\frac{-c^2\xi}{4}},\quad \tilde{H}=1+e^{\frac{-c^2\xi}{4}}\left(-\frac{1}{4c^2}+H\right).$$

These variables transform the equation (1.134) into its canonical form, i.e.,

$$\frac{\mathrm{d}^2\tilde{H}}{\mathrm{d}\tilde{\xi}^2}=\frac{1}{\tilde{\xi}}\left(256\left(\frac{\mathrm{d}\tilde{H}}{\mathrm{d}\tilde{\xi}}\right)^3-3\left(\frac{\mathrm{d}\tilde{H}}{\mathrm{d}\tilde{\xi}}\right)^2+3\frac{\mathrm{d}\tilde{H}}{\mathrm{d}\tilde{\xi}}-1\right),$$

that can be solved by two quadratures and, thus, its general solution is

$$\tilde{H}=\tilde{\xi}+c_2\pm 16\int\frac{\mathrm{d}\,\tilde{\xi}}{\sqrt{2c_1-2\log(\tilde{\xi})}}.$$

The last integral cannot be expressed in finite terms.
If we assume $c=0$ then (1.133) becomes

$$u_t=u_{xx}+x^2f(u).$$

In [32] the same nonclassical symmetry was determined if $f(u)=u^2(1-u)$ – (i) in(1.109). We found that this is true for any $f(u)$.

Also, we assume $k(x)=\dfrac{c}{e^{c(b_0-x)}-1}$. We solve the invariant surface equation (1.112), i.e.,

$$u_t+\frac{c}{e^{c(b_0-x)}-1}u_x=0,$$

and derive its complete solution as

$$u(t,x)=H(\gamma),\quad \gamma=-\frac{cx+c^2t+e^{c(b_0-x)}}{c^2},$$

where $H(\gamma)$ is an arbitrary function of γ. After substituting this solution into equation (1.107), i.e.,

$$u_t=u_{xx}+cu_x+\frac{\left(e^{c(b_0-x)}-1\right)^2}{c^2}f(u), \tag{1.135}$$

we obtain the following ordinary differential equation

$$\frac{\mathrm{d}^2H}{\mathrm{d}\gamma^2}+f(H)=0. \tag{1.136}$$

Its general solution in implicit form is

$$\pm \int \frac{\mathrm{d}\,H}{\sqrt{c_1 - 2\int f(H)\mathrm{d}\,H}} - \gamma - c_2 = 0.$$

Let us consider some instances:

(A) $f(u) = u^2 \implies u(t,x) = -6\text{WeierstrassP}(\gamma + c_1, 0, c_2)$,
where WeierstrassP represents the Weierstrass elliptic function.

(B) $f(u) = u^3 \implies u(t,x) = c_2\text{JacobiSN}\left(\left(\frac{\gamma}{\sqrt{2}} + c_1\right)c_2, \sqrt{-1}\right)$,

where $\text{JacobiSN}(z,k) = \sin(\text{JacobiAM}(z,k))$ and JacobiAM represents the Jacobi amplitude function AM.

(C) $f(u) = u^2(1-u) \implies \int \frac{6\,\mathrm{d}H}{\sqrt{18H^4 - 24H^3 + 36c_1}} - \gamma - c_2 = 0.$

Let us consider two particular values of c_1.

If we assume $c_1 = \frac{1}{6}$ then we obtain

$$u(t,x) = \frac{(1-H)\sqrt{18H^2+12H+6}}{\sqrt{18H^4-24H^3+6}}\text{arctanh}\left(\frac{2(1+2H)}{\sqrt{18H^2+12H+6}}\right) - \gamma - c_2 = 0,$$

although still an implicit solution of (1.135);
instead $c_1 = 0$ yields

$$u(t,x) = -\frac{12}{4c_2\gamma + 2c_2^2 + 2\gamma^2 - 9},$$

a nonclassical symmetry solution of equation (1.135), i.e.,

$$u_t = u_{xx} + cu_x + \frac{\left(e^{c(b_0-x)} - 1\right)^2}{c^2}u^2(1-u).$$

This solution tends to zero when t (or x) goes to infinity. Also it blows up in finite time if

$$t = \frac{1}{2c^2}\left(2c_2c^2 - 2cx - 2e^{c(b_0-x)} \pm c^2\sqrt{18}\right), \quad \forall x.$$

Moreover, we can assume in Case 2, as $k(x) = \frac{1}{b_1}\tan\left(\frac{x+b_2}{b_1}\right) - \frac{c}{2}$. Then, we solve the invariant surface equation (1.112), i.e.,

$$u_t + \left(\frac{1}{b_1}\tan\left(\frac{x+b_2}{b_1}\right) - \frac{c}{2}\right)u_x = 0,$$

and derive its complete solution as

$$
\begin{aligned}
&u(t,x) = H(\varrho),\\
&\quad \varrho = t + \frac{2b_1}{4+b_1^2c^2}\left[c(x+b_2) + \log\left(1+\tan^2\left(\frac{x+b_2}{b_1}\right)\right)\right.\\
&\quad \left. -2\log\left(2\tan\left(\frac{x+b_2}{b_1}\right) - b_1c\right)\right],
\end{aligned}
$$

where $H(\varrho)$ is an arbitrary function of ϱ. After substituting this solution into equation (1.107), i.e.,

$$
u_t = u_{xx} + cu_x + \frac{4b_1^2}{\left(2\tan\left(\frac{x+b_2}{b_1}\right) - b_1c\right)^2} f(u), \tag{1.137}
$$

we obtain the following ordinary differential equation

$$
4b_1^2\frac{\mathrm{d}^2H}{\mathrm{d}\varrho^2} + (4+b_1^2c^2)\frac{\mathrm{d}H}{\mathrm{d}\varrho} + 4b_1^2 f(H) = 0. \tag{1.138}
$$

If $f(u) = u$, namely if equation (1.137) is linear, then we obtain that the general solution is

$$
\begin{aligned}
u(t,x) &= c_1 e^{-\frac{4+b_1^2c^2 - \varrho\sqrt{(b_1^2c^2+4)^2 - 64b_1^4}}{8b_1^2}}\\
&+ c_2 e^{-\frac{4+b_1^2c^2 + \varrho\sqrt{(b_1^2c^2+4)^2 - 64b_1^4}}{8b_1^2}}.
\end{aligned}
$$

1.2.2 Nonclassical symmetries of the Black-Scholes equation

Now, we discuss the nonclassical symmetries of the *Black-Scholes* equation given by [72]

$$
u_t + \frac{1}{2}A^2x^2u_{xx} + Bxu_x - Bu = 0, \tag{1.139}
$$

where $A,\ B$ are arbitrary constants.

The G-equation of (1.139) is:

$$
G_t + BuG_u + A^2x\left(G_x + GG_u\right) + \frac{1}{2}A^2x^2G_{xx} = 0,
$$

and the η-equation is written as

$$
\begin{aligned}
&A^2x^2\left(G\eta\eta_{uG} + G\eta_{xu} + \eta\eta_{xG}\right) + \eta_t + B\eta + A^2\eta\\
&+ \frac{A^2x^2}{2}\left(G^2\eta_{uu} + \eta_{xx} + \eta^2\eta_{GG}\right) + Bx\eta_x + Bu\eta_u\\
&+ A^2x\left(\eta\eta_G + 2G\eta_u + 2\eta_x\right) = 0.
\end{aligned} \tag{1.140}
$$

The particular solution of the η-equation becomes

$$\eta(t,x,u,G)=\frac{2\left(V_2G-BxG+Bu-F\right)}{A^2x^2} \tag{1.141}$$

that replaced into (1.140) yields an overdetermined system in the unknowns F, V_2. Since we obtain a polynomial of third degree in G then we let MAPLE evaluate the four coefficients that we call $d_i,\;\; i=0,1,2,3$ where i stands for the corresponding power of G. We impose all of them to be zero. From $d_3=0$, we obtain

$$V_2(t,x,u)=\Theta_1(t,x)u+\Theta_2(t,x),$$

while d_2 yields

$$F(t,x,u)=\frac{1}{A^2x^2}\left(\frac{2\Theta_1^2u^3}{3}+2\Theta_1\Theta_2u^2-2B\Theta_1u^2\right)+\frac{\partial\Theta_1}{\partial x}u^2 \\ +\Theta_3(t,x)u+\Theta_4(t,x),$$

with $\Theta_j(t,x),\;\; j=1,\ldots,4$ arbitrary functions of t and x. Since d_1 is a polynomial of order 3 with respect to u we set $e_j,\;\; j=0,\ldots,3$ as the coefficients of $u^j,\;\; j=0,\ldots,3$. From $e_3=0$ we get

$$\Theta_1(t,x)=0, \tag{1.142}$$

which implies also $e_1=e_2=0$. Finally, we have

$$e_0=-3A^2x^3\Bigg[-x\Theta_2\left(B+4\frac{\partial\Theta_2}{\partial x}\right)+2Bx^2\frac{\partial\Theta_2}{\partial x}+A^3x^3\left(2\frac{\partial\Theta_3}{\partial x}-\frac{\partial^2\Theta_2}{\partial x^2}\right) \\ -2x\frac{\partial\Theta_2}{\partial t}+4\Theta_2^2\Bigg].$$

We consider the following special cases:

Case 1. $\Theta_2(t,x)=\tilde{\Theta}_2(t)$.

Case 2. $\Theta_2(t,x)=\tilde{\Theta}_2(x)$.

Case 1.

In this case by setting e_0 equals zero, one can obtain:

$$\tilde{\Theta}_2(t)=c_1e^{-Bt},\;\; \Theta_3(t,x)=\frac{c_1^2e^{-2Bt}}{A^2x^2}+\tilde{\Theta}_3(t).$$

One time differentiation of d_0 and setting it equal to zero yields

$$\tilde{\Theta}_3(t)=c_2,\;\; c_1=0.$$

By using these values we can write d_0 as follows:

$$d_0=-A^2x^3\left[A^2x^2\frac{\partial^2\Theta_4}{\partial x^2}+2\frac{\partial\Theta_4}{\partial t}+2Bx\frac{\partial\Theta_4}{\partial x}-2B\Theta_4\right]. \tag{1.143}$$

From $d_0 = 0$ some subcases are considerable.

Subcase 1.1. $\Theta_4(t,x) = \tilde{\Theta}_4(x)$
In this subcase by using $d_0 = 0$ we get

$$\tilde{\Theta}_4(x) = c_3 x + c_4 x^{\frac{-2B}{A^2}},$$

where c_3, c_4 are arbitrary constants. In this step all of the unknowns are determined and we have

$$V_2(t,x,u) = 0, \quad F(t,x,u) = c_2 u + c_3 x + c_4 x^{\frac{-2B}{A^2}},$$

and Eq. (1.141) becomes

$$\eta = \frac{2\left(-BxG + Bu - c_2 u - c_3 x - c_4 x^{\frac{-2B}{A^2}}\right)}{A^2 x^2},$$

namely

$$u_{xx} = \frac{2\left(-Bxu_x + Bu - c_2 u - c_3 x - c_4 x^{\frac{-2B}{A^2}}\right)}{A^2 x^2}. \tag{1.144}$$

Eq. (1.144) is a linear ordinary differential equation with respect to x and its solution is given by:

$$u(t,x) = \Psi_1(t) x^{\frac{A^2-2B+\lambda}{2A^2}} + \Psi_2(t) x^{\frac{A^2-2B-\lambda}{2A^2}} - \frac{c_3 x^2 + c_4 x^{\frac{A^2-2B}{A^2}}}{c_2 x}, \tag{1.145}$$

where $\lambda = \sqrt{A^4 + 4B^2 + 4BA^2 - 8c_2 A^2}$. Substituting (1.145) into (1.139) yields the following nonclassical symmetry solution

$$u(t,x) = c_5 e^{c_2 t} x^{\frac{A^2-2B+\lambda}{2A^2}} + c_6 e^{c_2 t} x^{\frac{A^2-2B-\lambda}{2A^2}} - \frac{c_3 x^2 + c_4 x^{\frac{A^2-2B}{A^2}}}{c_2 x},$$

with $c_k, k = 2, \ldots, 6$ arbitrary constants.

Subcase 1.2. $\Theta_4(t,x) = \tilde{\Theta}_4(t)$
From $d_0 = 0$ we get

$$\tilde{\Theta}_4(t) = c_3 e^{Bt}.$$

Therefore

$$V_2(t,x,u) = 0, \quad F(t,x,u) = c_2 u + c_3 e^{Bt},$$

and from (1.119), η-equation is as follows:

$$\eta = \frac{2\left(Bu - BxG - c_2 u - c_3 e^{Bt}\right)}{A^2 x^2},$$

which is equivalent to

$$u_{xx} = \frac{2\left(Bu - Bxu_x - c_2 u - c_3 e^{Bt}\right)}{A^2 x^2}.$$

Obtained equation is an ODE with respect to x and its solution is given by

$$u(t,x) = \Psi_1(t) x^{\frac{B+1-\lambda}{2}} + \Psi_2(t) x^{\frac{B+\lambda}{2}} + \frac{c_3 e^{Bt}}{B - c_2}, \tag{1.146}$$

where $\lambda = \sqrt{(B-1)^2 + 4c_2}$. Substituting (1.146) into (1.139) yields another nonclassical symmetry solution of (1.139) with

$$\Psi_1(t) = c_4 e^{-\frac{t\left((2+A^2)\left(B^2 - B\lambda - B\right) + 2A^2 c_2\right)}{4}}, \quad \Psi_2(t) = c_5 e^{-\frac{t\left((2+A^2)\left(B^2 + B\lambda - B\right) + 2A^2 c_2\right)}{4}},$$

where $c_k,\ k = 2, \ldots, 5$ are arbitrary constants and λ is defined as before.

Case 2.
In this case $e_0 = 0$ becomes as follows:

$$e_0 = 3A^2x^3\left[-2A^2x^3\frac{\partial\Theta_3}{\partial x} + A^2x^3\frac{d^2\tilde{\Theta}_2}{dx^2} + (4x\tilde{\Theta}_2 - 2Bx^2)\frac{d\tilde{\Theta}_2}{dx} - 4\tilde{\Theta}_2^2 + 2Bx\tilde{\Theta}_2\right] = 0. \tag{1.147}$$

To solve this equation, we consider the following subcases.

Subcase 2.1. $\Theta_3(t,x) = \frac{1}{2}\frac{d\tilde{\Theta}_2(x)}{dx}$
Eq. (1.147) yields $\tilde{\Theta}_2(x) = c_1 x$ and therefore (1.143) becomes

$$d_0 = x\left[A^2x^2\frac{\partial^2\Theta_4}{\partial x^2} + 2Bx\frac{\partial\Theta_4}{\partial x} + 2\frac{\partial\Theta_4}{\partial t} - 2B\Theta_4\right]. \tag{1.148}$$

By setting $\Theta_4(t,x) = \tilde{\Theta}_4(x)$ in (1.148), and solving $d_0 = 0$ we get

$$\tilde{\Theta}_4(x) = A_1 x + A_2 x^{-\frac{2B}{A^2}}.$$

Therefore

$$V_2(t,x,u) = c_1 x, \quad F(t,x,u) = \frac{1}{2}c_1 u + A_1 x + A_2 x^{-\frac{2B}{A^2}},$$

and η-equation becomes

$$\eta = \frac{2(c_1 - B)xG + 2Bu - c_1 u - 2A_1 x - 2A_2 x^{-\frac{2B}{A^2}}}{A^2x^2};$$

in other words:

$$u_{xx} = \frac{2(c_1 - B)xu_x + 2Bu - c_1 u - 2A_1 x - 2A_2 x^{-\frac{2B}{A^2}}}{A^2x^2}. \tag{1.149}$$

Eq. (1.144) is obtainable from Eq. (1.149) by setting $c_1 = \frac{B}{2} = 2c_2$. Thus we escape from the calculation of solutions for this case.
However, setting $\Theta_4(t,x) = \tilde{\Theta}_4(t)$ in (1.148), and solving $d_0 = 0$ yields

$$\tilde{\Theta}_4(t) = A_1 e^{Bt}.$$

Therefore

$$V_2(t,x,u) = c_1 x, \quad F(t,x,u) = \frac{1}{2}c_1 u + A_1 e^{Bt}.$$

Thus, η-equation becomes

$$\eta = \frac{2(c_1 - B)xG + 2Bu - c_1 u - 2A_1 e^{Bt}}{A^2 x^2};$$

in other words:

$$u_{xx} = \frac{2(c_1 - B)xu_x + (2B - c_1)u - 2A_1 e^{Bt}}{A^2 x^2}. \tag{1.150}$$

Solving the obtained ODE yields:

$$u(t,x) = \Psi_1(t)x^{\frac{2c_1 - 2B + A^2 - \lambda}{2A^2}} + \Psi_2(t)x^{\frac{2c_1 - 2B + A^2 + \lambda}{2A^2}} + \frac{2A_1 e^{Bt}}{2B - c_1}, \tag{1.151}$$

where $\lambda = \sqrt{4(c_1 - B)^2 + 4A^2 B + A^4}$. Another nonclassical symmetry solution of (1.139) is obtainable as follows, by substituting (1.151) into (1.139)

$$\begin{aligned}
\Psi_1(t) &= A_2 e^{c_1 t\left(\frac{2B - 2c_1 + \lambda}{2A^2}\right)},\\
\Psi_2(t) &= A_3 e^{c_1 t\left(\frac{2B - 2c_1 + \lambda}{2A^2}\right)} + e^{c_1 t\left(\frac{2B - 2c_1 + \lambda}{2A^2}\right)}\\
&\times \frac{4A_1 B(2B - 2c_1 + \lambda)x^{\frac{2B - 2c_1 - \lambda - A^2}{2A^2}} e^{t\left(\frac{2BA^2 - 2Bc_1 + 2c_1^2 + c_1\lambda}{2A^2}\right)}}{c_1(4B + A^2)(2B - c_1)},
\end{aligned}$$

where $A_k,\ k = 1,2,3$ and c_1 are arbitrary constants and λ is defined as before.

Subcase 2.2. $\Theta_3(t,x) = \tilde{\Theta}_3(x)$
From Eq. (1.147) we have

$$\tilde{\Theta}_3(x) = \frac{\tilde{\Theta}_2(x)^2 - B\tilde{\Theta}_2(x)}{A^2 x^2} + \frac{1}{2}\frac{d\tilde{\Theta}_2(x)}{dx} + A_1.$$

By this value of $\tilde{\Theta}_3(x)$, equation $d_0 = 0$ becomes a polynomial of order one with respect to u. A special solution of coefficient of u in $d_0 = 0$ is

$$\tilde{\Theta}_2(x) = \frac{x\left(A^2 + \sqrt{A^4 + 8A^2 B + 8B^2 - 16A^2 A_1}\right)}{4}.$$

Thus $d_0 = 0$ becomes an equation without u, which by setting $\Theta_4(t,x) = \tilde{\Theta}_4(t)$ we get

$$\tilde{\Theta}_4(t) = k_1 e^{Bt}.$$

Therefore

$$V_2(t,x,u) = \frac{(A^2+\lambda)x}{4},$$
$$F(t,x,u) = \frac{(A^2B + A^4 + 2B^2)u - \lambda(Bu - A^2u^2) + 4k_1A^2e^{Bt}}{4A^2},$$

where $\lambda = \sqrt{A^4 + 8A^2B + 8B^2 - 16A_1A^2}$ and η-equation becomes

$$\eta = \frac{A^4xG + \lambda(A^2xG + Bu - A^2u) - 4A^2BxG}{2A^4x^2} + \frac{3A^2Bu - A^4u - 2B^2u - 4k_1A^2e^{Bt}}{2A^4x^2},$$

or equivalently

$$u_{xx} = \frac{A^4xu_x + \lambda(A^2xu_x + Bu - A^2u) - 4A^2Bxu_x}{2A^4x^2} + \frac{3A^2Bu - A^4u - 2B^2u - 4k_1A^2e^{Bt}}{2A^4x^2}. \tag{1.152}$$

Another nonclassical symmetry solution of Eq. (1.139) can be found by solving Eq. (1.152) which is as follows:

$$u(t,x) = \Psi_1(t)x^{\frac{A^2-2B+\lambda}{2A^2}} + \Psi_2(t)x^{\frac{A^2-B}{A^2}} - \frac{4k_1A^2e^{Bt}}{(A^2-B)(A^2-2B+\lambda)},$$

where

$$\Psi_1(t) = e^{t\left(\frac{4A_1A^2-A^2B-B^2}{2A^2}\right)}, \quad \Psi_2(t) = e^{tB\left(\frac{A^2+B}{2A^2}\right)},$$

and $\lambda = \sqrt{A^4 + 8A^2B + 8B^2 - 16A_1A^2}$.
Also, if in $d_0 = 0$ we set $\Theta_4(t,x) = \tilde{\Theta}_4(x)$, then

$$\tilde{\Theta}_4(x) = k_1x + k_2x^{-\frac{2B}{A^2}},$$

and therefore

$$V_2(t,x,u) = \frac{x\left(A^2+\lambda\right)}{4},$$
$$F(t,x,u) = \frac{A^2Bx^2u - \lambda(Bx^2 - A^2X^2)u + A^4x^2u}{4A^2x^2} + \frac{4k_1A^2x^3 + 2B^2x^2u + 4k_2A^2x^{2\left(\frac{A^2-B}{A^2}\right)}}{4A^2x^2},$$

where $\lambda = \sqrt{A^4 + 8A^2B + 8B^2 - 16A_1A^2}$. Hence, the η-equation is as follows:

$$\eta = \frac{A^4x^3G + \lambda(A^2x^3G + Bux^2 - A^2ux^2) - 4A^2Bx^3G}{2A^4x^4} + \frac{3A^2Bux^2 - A^4ux^2 - 2B^2ux^2 - 4k_1A^2x^3 - 4k_2A^2x^{2\left(\frac{A^2-B}{A^2}\right)}}{2A^4x^2};$$

in other words

$$u_{xx} = \frac{A^4x^3u_x + \lambda(A^2x^3u_x + Bux^2 - A^2ux^2) - 4A^2Bx^3u_x}{2A^4x^4} + \frac{3A^2Bux^2 - A^4ux^2 - 2B^2ux^2 - 4k_1A^2x^3 - 4k_2A^2x^{\frac{2A^2-2B}{A^2}}}{2A^4x^2},$$

from which its solution is given by

$$\begin{aligned} u(t,x) &= \Psi_1(t)x^{\frac{A^2-2B+\lambda}{2A^2}} + \Psi_2(t)x^{\frac{A^2-B}{A^2}} \\ &+ \frac{2A^2(A^2+B)(k_1x^{\frac{A^2+2B}{A^2}}\mu - \lambda B(k_1x^{\frac{A^2+2B}{A^2}} + k_2))}{x^{\frac{2B}{A^2}}(4A_1A^2 - A^2B - B^2)(\lambda - A^2)(A^2B + B^2)} \\ &+ \frac{k_2B(5A^2B + 4B^2 + A^4 - 8A_1A^2)}{x^{\frac{2B}{A^2}}(4A_1A^2 - A^2B - B^2)(\lambda - A^2)(A^2B + B^2)}, \end{aligned}$$

where $\mu = 8A^2A_1 - 3A^2B - 4B^2$ and similar to previous

$$\Psi_1(t) = k_3e^{t\left(\frac{4A_1A^2 - A^2B - B^2}{2A^2}\right)}, \quad \Psi_2(t) = k_4e^{tB\left(\frac{A^2+B}{2A^2}\right)}.$$

Comparison of presented solutions of Eq. (1.139) in literature with non-classical solutions shows that reported solutions in this section are new.

1.3 Self-adjointness and conservation laws

In this section, after some preliminaries, conservation laws by using the new conservation theorem introduced in [95] will be considered for different problems.

Consider a kth-order PDE of n independent variables $x = (x^1, x^2, \dots, x^n)$ and dependent variable u, viz.,

$$F(x, u, u, u_{(1)}, \dots, u_{(k)}) = 0, \tag{1.153}$$

where $u_{(1)} = \{u_i\},\ u_{(2)} = \{u_{ij}\}, \dots$ and $u_i = \mathcal{D}_i(u),\ u_{ij} = \mathcal{D}_j\mathcal{D}_i(u)$, where

$$\mathcal{D}_i = \frac{\partial}{\partial x^i} + u_i\frac{\partial}{\partial u} + u_{ij}\frac{\partial}{\partial u_j} + \cdots, \quad i = 1, 2, \dots, n,$$

are the total derivative operators with respect to x^is.
The Euler-Lagrange operator, by formal sum, is given by

$$\frac{\delta}{\delta u} = \frac{\partial}{\partial u} + \sum_{s\geq 1}(-1)^s \mathcal{D}_{i_1}\cdots\mathcal{D}_{i_s}\frac{\partial}{\partial u_{i_1\cdots i_s}}. \tag{1.154}$$

Also, if $\mathcal{A}$ be the set of all differential functions of all finite orders, and $\xi^i, \eta \in \mathcal{A}$, then *Lie-Bäcklund* operator is

$$X = \xi^i\frac{\partial}{\partial x^i} + \eta\frac{\partial}{\partial u} + \zeta_i\frac{\partial}{\partial u_i} + \zeta_{i_1i_2}\frac{\partial}{\partial u_{i_1i_2}} + \cdots, \tag{1.155}$$

where

$$\zeta_i = \mathcal{D}_i(\eta) - u_j\mathcal{D}_i(\xi^j),$$
$$\zeta_{i_1\ldots i_s} = \mathcal{D}_{i_s}(\zeta_{i_1\ldots i_{s-1}}) - u_{ji_1\ldots i_{s-1}}\mathcal{D}_{i_s}(\xi^j), \quad s>1.$$

One can write the *Lie-Bäcklund* operator (1.155) in characteristic form

$$X = \xi^i\mathcal{D}_i + W\frac{\partial}{\partial u} + \sum_{s\geq 1}\mathcal{D}_{i_1}\ldots\mathcal{D}_{i_s}(W)\frac{\partial}{\partial u_{i_1i_2\ldots i_s}},$$

where

$$W = \eta - \xi^j u_j \tag{1.156}$$

is the characteristic function.
Euler-Lagrange operators with respect to derivatives of u are obtained by replacing u and the corresponding derivatives in (1.154), e.g.,

$$\frac{\delta}{\delta u_i} = \frac{\partial}{\partial u_i} + \sum_{s\geq 1}(-1)^s \mathcal{D}_{j_1}\cdots\mathcal{D}_{j_s}\frac{\partial}{\partial u_{ij_1\cdots j_s}}. \tag{1.157}$$

There is a connection between the Euler-Lagrange, *Lie-Bäcklund* and the associated operators by the following identity:

$$X + \mathcal{D}_i(\xi^i) = W\frac{\delta}{\delta u} + \mathcal{D}_i\mathcal{N}^i,$$

where

$$\mathcal{N}^i = \xi^i + W\frac{\delta}{\delta u_i} + \sum_{s\geq 1}\mathcal{D}_{i_1}\cdots\mathcal{D}_{i_s}(W)\frac{\delta}{\delta u_{ii_1\cdots i_s}}, \quad i = 1,\ldots,n,$$

are the Noether operators associated with a *Lie-Bäcklund* symmetry operator. The n-tuple vector $T = (T^1, T^2, \ldots, T^n)$, $T^i \in \mathcal{A}, i = 1,\ldots,n$, is a conserved vector of Eq. (1.153) if

$$\mathcal{D}_i(T^i) = 0 \tag{1.158}$$

on the solution space of (1.153). The expression (1.158) is a local conservation law of Eq. (1.153) and $T^i \in \mathcal{A}$ are called the fluxes of the conservation law.

Definition 7 *A local conservation law (1.158) of the PDE (1.153) is trivial if its fluxes are of the form $T^i = M^i + H^i$, where M^i and H^i are functions of x, u and derivatives of u such that M^i vanishes on the solutions of the system (1.153), and $\mathcal{D}_i H^i = 0$ is identically divergence-free.*

In particular, a trivial conservation law contains no information about a given PDE (1.153) and arises in two cases:
1. Each of its fluxes vanishes identically on the solutions of the given PDE.
2. The conservation law vanishes identically as a differential identity. In particular, this second type of trivial conservation law is simply an identity holding for arbitrary fluxes. These $T = (T^1, T^2, ..., T^n)$ are called null divergences.

The adjoint equation to the kth-order differential Eq. (1.153) is defined by

$$F^*(x, u, v, u_{(1)}, v_{(1)}, \dots, u_{(k)}, v_{(k)}) = 0, \tag{1.159}$$

where

$$F^*(x, u, v, u_{(1)}, v_{(1)}, \dots, u_{(k)}, v_{(k)}) = \frac{\delta(v^\beta F_\beta)}{\delta u}, \quad v = v(x),$$

and $v = (v^1, v^2, \dots, v^m)$ are new dependent variables.
We recall here the following results as given in Ibragimov's paper [95].

Definition 8 *[95] Eq. (1.153) is said to be self-adjoint if the substitution of $v = u$ into adjoint Eq. (1.159) yields the same Eq. (1.153).*

Definition 9 *[96] Eq. (1.153) is said to be quasi self-adjoint if the equation obtained from the adjoint Eq. (1.159) by the substitution $v = h(u)$, with a certain function $h(u)$ such that $h'(u) \neq 0$ is identical to the original equation.*

Definition 10 *[61] Eq. (1.153) is said to be weakly self-adjoint if the equation obtained from the adjoint Eq. (1.159) by the substitution $v = h(t, x, u)$, with a certain function $h(t, x, u)$ such that $h_t(t, x, u) \neq 0$, (or $h_x(t, x, u) \neq 0$) and $h_u(t, x, u) \neq 0$ is identical to the original equation.*

Definition 11 *[98] Eq. (1.153) is said to be nonlinearly self-adjoint if the equation obtained from the adjoint Eq. (1.159) by the substitution $v = h(x, u, u_{(1)}, \dots)$, with a certain function $h(x, u, u_{(1)}, \dots)$ such that $h(x, u, u_{(1)}, \dots) \neq$ constant is identical to the original equation (1.153).*

The main theorem which is used to construct the conservation laws is given as follows:

Theorem 1 *[95] Every Lie point, Lie-Bäcklund and nonlocal symmetry admitted by the Eq. (1.153) gives rise to a conservation law for the system consisting of the Eq. (1.153) and the adjoint Eq. (1.159) where the components T^i of the conserved vector $T = (T^1, \dots, T^n)$ are determined by*

$$T^i = \xi^i \mathcal{L} + W \frac{\delta \mathcal{L}}{\delta u_i} + \sum_{s \geq 1} \mathcal{D}_{i_1} \dots \mathcal{D}_{i_s}(W) \frac{\delta \mathcal{L}}{\delta u_{i i_1 i_2 \dots i_s}}, \quad i = 1, \dots n, \tag{1.160}$$

with Lagrangian given by

$$\mathcal{L} = vF(x, u, \ldots, u_{(k)}).$$

1.3.1 Conservation laws of the Black-Scholes equation

Here, we obtain the conservation laws of the *Black-Scholes* equation (1.139). The adjoint equation for Eq. (1.139) is as follows:

$$F^* = \frac{\delta(vF)}{\delta u} = \frac{\delta\left(v\left[u_t + \frac{1}{2}A^2x^2u_{xx} + Bxu_x - Bu\right]\right)}{\delta u},$$

which by some simplifications we get

$$F^* = -2Bv - v_t - Bxv_x + A^2v + 2A^2xv_x + \frac{1}{2}A^2x^2v_{xx} = 0. \tag{1.161}$$

By setting $t = x^1$ and $x = x^2$, the conservation law will be written

$$\mathcal{D}_t(T_i^t) + \mathcal{D}_x(T_i^x) = 0, \quad i = 1, \ldots, 7.$$

Now, we discuss self-adjointness of Eq. (1.139) by the following theorem.

Theorem 2 *Eq. (1.139) is neither quasi self-adjoint nor weakly self-adjoint; however Eq. (1.139) is nonlinearly self-adjoint for*

$$h(t, x, u) = e^{c_1 t}\left(c_2 x^{\frac{2B-3A^2+\chi}{2A^2}} + c_3 x^{\frac{2B-3A^2-\chi}{2A^2}}\right), \tag{1.162}$$

where $\chi = \sqrt{4B^2 + 4BA^2 + A^4 + 8A^2c_1}$.

Proof: By a few computations we can show that Eq. (1.139) is neither quasi self-adjoint nor weakly self-adjoint. To demonstrate the nonlinear self-adjointness, setting $v = h(t, x, u)$ in Eq. (1.167) we get

$$\begin{aligned}
&-2Bh + A^2h - Bx(h_x + h_u u_x) + 2A^2x\,(h_x + h_u u_x) - h_t - h_u u_t \\
&+ \frac{1}{2}A^2x^2(h_{xx} + 2h_{xu}u_x + h_{uu}u_x^2 + h_u u_{xx}) = 0,
\end{aligned}$$

which yields:

$$\begin{aligned}
F^* - \lambda\left(u_t + \frac{1}{2}A^2x^2u_{xx} + Bxu_x - Bu\right) &= -\lambda u_t - \lambda Bxu_x + \lambda Bu - 2hB \\
- h_t + A^2h - h_u u_t + \frac{1}{2}A^2x^2(h_{xx} &+ h_{uu}u_x^2 + h_u u_{xx} - \lambda u_{xx} + 2h_{xu}u_x) \\
- Bxh_x - Bxh_u u_x + 2A^2x(h_x + h_u u_x) &= 0.
\end{aligned} \tag{1.163}$$

Comparing the coefficients for the different derivatives of u we obtain some conditions, one of which is $\lambda + h_u = 0$. Thus, by setting $\lambda = -h_u$ in (1.169) we get

$$A^2x^2(h_u u_{xx} + h_{xu}u_x) - Bh_u u - 2Bh - h_t - Bxh_x + A^2h + 2A^2x(h_x + h_u u_x) + \frac{1}{2}A^2x^2(h_{xx} + h_{uu}u_x^2) = 0. \quad (1.164)$$

As previously, comparing the coefficients for the different derivatives of u, we have the following condition:

$$\frac{A^2x^2}{2}h_{xx} + (2A^2 - B)xh_x + (A^2 - 2B)h - h_t = 0, \quad (1.165)$$

which by solving this system completes the proof.

Here infinite dimensional Lie algebras of Eq. (1.139) presented in [121] are used to construct the infinite number of conservation laws.

Eq. (1.139) admits six-dimensional Lie algebras; thus we consider the following seven cases:

(i) We first consider the Lie point symmetry generator $X_1 = \frac{\partial}{\partial t}$. The components of the conserved vector are given by

$$T_1^t = \frac{1}{2}A^2x^2vu_{xx} + Bxvu_x - Bvu,$$
$$T_1^x = -Bxvu_t + A^2xvu_t + \frac{1}{2}A^2x^2(u_tv_x - vu_{tx}).$$

By setting $c_1 = c_3 = 0$ and $c_2 = 1$ in Theorem 9, we get

$$T_1^t|_{v=\frac{1}{x^2}} = \frac{A^2}{2}u_{xx} + \frac{B}{x}u_x - \frac{B}{x^2}u = \mathcal{D}_x\left(\frac{A^2}{2}u_x + \frac{B}{x}u\right),$$
$$T_1^x|_{v=\frac{1}{x^2}} = -\frac{A^2}{2}u_{tx} - \frac{B}{x}u_t = -\mathcal{D}_t\left(\frac{A^2}{2}u_x + \frac{B}{x}u\right).$$

Then transferring the terms $\mathcal{D}_x(\cdots)$ from T_1^t to T_1^x provides the null divergence $T_1 = (T_1^t, T_1^x) = (0, 0)$.

(ii) Using the Lie point symmetry generator $X_2 = x\frac{\partial}{\partial x}$, the components of the conserved vector are given by

$$T_2^t = -xvu_x,$$
$$T_2^x = xvu_t - Bxvu + \frac{1}{2}A^2x^2(vu_x + xu_xv_x).$$

Setting $v = h(t, x, u) = \frac{1}{x^2}$ into T_2^t, T_2^x and after reckoning, we have

$$T_2^t|_{v=\frac{1}{x^2}} = -\frac{u}{x^2} + \mathcal{D}_x\left(-\frac{u}{x}\right),$$
$$T_2^x|_{v=\frac{1}{x^2}} = -\frac{B}{x}u - \frac{A^2}{2}u_x - \mathcal{D}_t\left(-\frac{u}{x}\right).$$

Therefore $T_2 = (T_2^t, T_2^x) = (-\frac{u}{x^2}, -\frac{B}{x}u - \frac{A^2}{2}u_x)$.

(iii) Using the Lie point symmetry generator $X_3 = u\frac{\partial}{\partial u}$, one can obtain the conserved vector whose components are

$$
\begin{aligned}
T_3^t =& uv, \\
T_3^x =& \frac{1}{2}A^2x^2(vu_x - uv_x) + Bxvu - A^2xuv.
\end{aligned}
$$

Setting $v = \frac{1}{x^2}$ concludes the previous conserved vectors; however $c_1 = c_2 = 0$ and $c_3 = 1$ in (1.168) yields

$$
\begin{aligned}
T_3^t\Big|_{v=x^{\left(\frac{2B}{A^2}-1\right)}} =& x^{\left(\frac{2B}{A^2}-1\right)}u, \\
T_3^x\Big|_{v=x^{\left(\frac{2B}{A^2}-1\right)}} =& \frac{A^2}{2}\left(x^{\left(\frac{2B}{A^2}+1\right)}u_x - x^{\left(\frac{2B}{A^2}\right)}u\right).
\end{aligned}
$$

(iv) Utilizing the Lie point symmetry generator $X_4 = 2tx\frac{\partial}{\partial x} + \left(tu - \frac{2Btu}{A^2} + \frac{2u\ln(x)}{A^2}\right)\frac{\partial}{\partial u}$, one can obtain the conserved vector whose components are

$$
\begin{aligned}
T_4^t &= \left(A^2tu - 2Btu + 2u\ln(x) - 2A^2txu_x\right)v, \\
T_4^x &= -\frac{1}{2}x\Big(- 4A^2tvu_t + 2A^2Btxvu_x - 2A^2Btvu - 2A^2vu - 2BA^2xtuv_x \\
&+ A^4t(2uv + xuv_x - 2x^2u_xv_x - 3xu_xv) + 4B^2tuv \\
&+ \ln(x)(4A^2uv - 4Buv + 2A^2xuv_x - 2A^2xvu_x)\Big).
\end{aligned}
$$

Substituting $v = \frac{1}{x^2}$ into the components above, we obtain

$$
\begin{aligned}
T_4^t|_{v=\frac{1}{x^2}} &= \left(\frac{2\ln(x) - A^2t - 2Bt}{x^2}\right)u + \mathcal{D}_x\left(\frac{-2A^2tu}{x}\right), \\
T_4^x|_{v=\frac{1}{x^2}} &= \frac{-2A^2Btxu_x - 2A^2Btu - 4B^2tu + 4Bu\ln(x) - A^4txu_x}{2x} \\
&+ \frac{2A^2x\ln(x)u_x - 2A^2u}{2x} - \mathcal{D}_t\left(\frac{-2A^2tu}{x}\right).
\end{aligned}
$$

Then transferring the terms $\mathcal{D}_x(\cdots)$ from T_4^t to T_4^x provides

$$
\begin{aligned}
T_4^t|_{v=\frac{1}{x^2}} &= \left(\frac{2\ln(x) - A^2t - 2Bt}{x^2}\right)u, \\
T_4^x|_{v=\frac{1}{x^2}} &= \frac{-2A^2Btxu_x - 2A^2Btu - 4B^2tu + 4Bu\ln(x) - A^4txu_x}{2x} \\
&+ \frac{2A^2x\ln(x)u_x - 2A^2u}{2x}.
\end{aligned}
\tag{1.166}
$$

(v) Using the Lie point symmetry generator

$$X_5 = 8t\frac{\partial}{\partial t} + 4x\ln(x)\frac{\partial}{\partial x} + \left(A^2tu + 4Btu + \frac{4B^2tu}{A^2} + 2u\ln(x) - \frac{4Bu\ln(x)}{A^2}\right)\frac{\partial}{\partial u},$$

one can obtain the conserved vector whose components are

$$T_5^t = v\Big(4A^4tx^2u_{xx} + 8A^2Btxu_x - 4A^2Btu + A^4tu + 4B^2tu + \ln(x)(2A^2u - 4Bu - 4A^2xu_x)\Big),$$

$$T_5^x = -\frac{1}{2}x\Big(6A^4Btuv + A^6tx(uv_x - vu_x) + 16A^2Btvu_t + 4B^2A^2txuv_x + 2A^6tuv + A^4tx(8vu_{tx} - 8u_tv_x - 4Bvu_x + 4Buv_x) + A^4\ln(x)(2xuv_x - 6xvu_x - 4x^2u_xv_x + 4uv) - 4A^2Bvu\ln(x) - 4A^2B^2txvu_x + 4A^2Bx\ln(x)(vu_x - uv_x) - 8A^2\ln(x)vu_t - 16A^4tu_tv - 8B^3tuv + 4A^2Bvu + 4A^4xvu_x + 8B^2u\ln(x)v - 2A^4vu\Big),$$

which by setting $v = \frac{1}{x^2}$ we get

$$T_5^t|_{v=\frac{1}{x^2}} = \frac{4A^2Bt + 4A^2 - 2A^2\ln(x) + A^4t + 4B^2t - 4B\ln(x)}{x^2}u + D_x\left(4A^4tu_x + (8A^2Bt - 4A^2\ln(x))\frac{u}{x}\right),$$

$$T_5^x|_{v=\frac{1}{x^2}} = \frac{1}{2x}\Big(12A^2Bu + 4A^4xu_x + A^6txu_x + 2A^4Btu - 4A^2B\ln(x)u - 2A^4x\ln(x)u_x + 8A^2B^2tu + 4A^4Btxu_x + 4A^2B^2txu_x - 4A^2Bx\ln(x)u_x + 2A^4u + 8B^3tu - 8B^2\ln(x)u\Big) - D_t\left(4A^4tu_x + (8A^2Bt - 4A^2\ln(x))\frac{u}{x}\right);$$

therefore

$$T_5^t|_{v=\frac{1}{x^2}} = \frac{4A^2Bt + 4A^2 - 2A^2\ln(x) + A^4t + 4B^2t - 4B\ln(x)}{x^2}u,$$

$$T_5^x|_{v=\frac{1}{x^2}} = \frac{1}{2x}\Big(12A^2Bu + 4A^4xu_x + A^6txu_x + 2A^4Btu - 4A^2B\ln(x)u - 2A^4x\ln(x)u_x + 8A^2B^2tu + 4A^4Btxu_x + 4A^2B^2txu_x - 4A^2Bx\ln(x)u_x + 2A^4u + 8B^3tu - 8B^2\ln(x)u\Big).$$

(vi) Using the Lie point symmetry generator

$$X_6 = 8t^2\frac{\partial}{\partial t} + 8tx\ln(x)\frac{\partial}{\partial x} + \left(A^2t^2u - 4tu + 4Bt^2u + 4tu\ln(x) + \frac{4B^2t^2u}{A^2} - \frac{8Btu\ln(x)}{A^2} + \frac{4u\ln^2(x)}{A^2}\right)\frac{\partial}{\partial u},$$

one can obtain the conserved vector whose components are

$$T_6^t = v\Big(4A^4t^2x^2u_{xx} + 8A^2Bt^2xu_x - 4A^2Bt^2u + A^4t^2u - 4A^2tu + 4B^2t^2u + \ln(x)(4A^2tu - 8Btu - 8A^2txu_x) + 4u\ln(x)^2\Big),$$

$$\begin{aligned}T_6^x = -\frac{1}{2}x\Big(&16A^2Btvu - 4A^4txuv_x + 12A^4txu_xv + 8A^4tu\ln(x)v - 16A^2t\ln(x)vu_t\\ &+A^4t^2(6Buv - 8xu_tv_x + 8xvu_{tx} - 16u_tv - 4Bxvu_x) + 16A^2Bt^2u_tv - 4A^2B^2xt^2vu_x\\ &+\ln(x)^2(4A^2xuv_x + 8A^2uv - 4A^2xvu_x - 8Buv) + A^6t^2xuv_x + 16B^2tu\ln(x)v\\ &+\ln(x)(-8A^2Btvu + 4A^4txuv_x - 12A^4txu_xv) - 8A^4tx^2\ln(x)u_xv_x + 4A^4Bxt^2uv_x\\ &+4A^2B^2t^2xuv_x + \ln(x)(8A^2Bxtvu_x - 8A^2Bxtuv_x - 8A^2uv) + 2A^6t^2uv\\ &-12A^4tuv - 8B^3t^2uv - A^6t^2xvu_x\Big).\end{aligned}$$

Setting $v = \frac{1}{x^2}$ into the components of T_6^t and T_6^x we get

$$\begin{aligned}T_6^t|_{v=\frac{1}{x^2}} &= \frac{4A^2Bt^2 + A^4t^2 + 4A^2t + 4B^2t^2 - 4A^2t\ln(x) - 8Bt\ln(x) + 4\ln^2(x)}{x^2}u\\ &\quad + \mathcal{D}_x\left(4A^4t^2u_x + (8A^2Bt^2 - 8A^2t\ln(x))\frac{u}{x}\right),\\ T_6^x|_{v=\frac{1}{x^2}} &= \frac{1}{2x}\Big(-8A^2Btx\ln(x)u_x + A^6t^2xu_x + 16A^2Btu + 4A^4txu_x + 2A^4Bt^2u\\ &\quad - 16B^2t\ln(x)u + 4A^2x\ln^2(x)u_x + 8A^2B^2t^2u + 4A^4Bt^2xu_x + 4A^2B^2t^2xu_x\\ &\quad - 8A^2Bt\ln(x)u - 4A^4tx\ln(x)u_x + 4A^4tu - 8A^2\ln(x)u + 8B^3t^2u\\ &\quad + 8B\ln^2(x)u\Big) - \mathcal{D}_t\left(4A^4t^2u_x + (8A^2Bt^2 - 8A^2t\ln(x))\frac{u}{x}\right);\end{aligned}$$

thus

$$\begin{aligned}T_6^t|_{v=\frac{1}{x^2}} &= \frac{4A^2Bt^2 + A^4t^2 + 4A^2t + 4B^2t^2 - 4A^2t\ln(x) - 8Bt\ln(x) + 4\ln^2(x)}{x^2}u,\\ T_6^x|_{v=\frac{1}{x^2}} &= \frac{1}{2x}\Big(-8A^2Btx\ln(x)u_x + A^6t^2xu_x + 16A^2Btu + 4A^4txu_x + 2A^4Bt^2u\\ &\quad - 16B^2t\ln(x)u + 4A^2x\ln^2(x)u_x + 8A^2B^2t^2u + 4A^4Bt^2xu_x + 4A^2B^2t^2xu_x\\ &\quad - 8A^2Bt\ln(x)u - 4A^4tx\ln(x)u_x + 4A^4tu - 8A^2\ln(x)u + 8B^3t^2u\\ &\quad + 8B\ln^2(x)u\Big).\end{aligned}$$

(vii) Using the Lie point symmetry generator $X_7 = \varphi(t,x)\frac{\partial}{\partial u}$ where $\varphi(t,x)$ satisfies the following equation:

$$2\varphi_t - 2B\varphi + 2Bx\varphi_x + A^2x^2\varphi_{xx} = 0,$$

one can obtain the conserved vector whose components are

$$T_7^t = v\varphi,$$
$$T_7^x = \frac{1}{2}x\Big((2Bv - 2A^2v - A^2xv_x)\varphi + A^2xv\varphi_x\Big).$$

Substituting $v = \frac{1}{x^2}$ into the components above, we obtain

$$T_7^t|_{v=\frac{1}{x^2}} = \frac{\varphi}{x^2},$$
$$T_7^x|_{v=\frac{1}{x^2}} = \frac{2B\varphi + A^2x\varphi_x}{2x}.$$

Since

$$\mathcal{D}_t\left(\frac{\varphi}{x^2}\right) + \mathcal{D}_x\left(\frac{2B\varphi + A^2x\varphi_x}{2x}\right) = 0,$$

it follows that the vector $T_7 = (T_7^t, T_7^x)$ is a local conserved vector for equation (1.139).

1.3.2 Conservation laws of the couple stress fluid-filled thin elastic tubes

The adjoint equation for the Eq. (1.66) is as follows:

$$F^* = \frac{\delta(vF)}{\delta u} = \frac{\delta\big(v\big[u_\tau + a_1uu_\xi - a_2u_{\xi\xi} + a_3u_{\xi\xi\xi} + a_4u_{\xi\xi\xi\xi}\big]\big)}{\delta u},$$

which by some simplifications we get

$$F^* = -v_\tau - a_1v_\xi u - a_2v_{\xi\xi} - a_3v_{\xi\xi\xi} + a_4v_{\xi\xi\xi\xi} = 0. \tag{1.167}$$

By setting $\tau = x^1$ and $\xi = x^2$, the conservation law will be written

$$\mathcal{D}_\tau(T_i^\tau) + \mathcal{D}_\xi(T_i^\xi) = 0, \quad i = 1, \ldots, 3,$$

and

$$T_i^\tau = \zeta^1\mathcal{L} + W\left[\frac{\partial\mathcal{L}}{\partial u_\tau}\right],$$
$$T_i^\xi = \zeta^2\mathcal{L} + W\left[\frac{\partial\mathcal{L}}{\partial u_\xi} - \mathcal{D}_\xi\left(\frac{\partial\mathcal{L}}{\partial u_{\xi\xi}}\right) + \mathcal{D}_\xi^2\left(\frac{\partial\mathcal{L}}{\partial u_{\xi\xi\xi}}\right) - \mathcal{D}_\xi^3\left(\frac{\partial\mathcal{L}}{\partial u_{\xi\xi\xi\xi}}\right)\right]$$
$$+ \mathcal{D}_\xi(W)\left[\frac{\partial\mathcal{L}}{\partial u_{\xi\xi}} - \mathcal{D}_\xi\left(\frac{\partial\mathcal{L}}{\partial u_{\xi\xi\xi}}\right) + \mathcal{D}_\xi^2\left(\frac{\partial\mathcal{L}}{\partial u_{\xi\xi\xi\xi}}\right)\right]$$
$$+ \mathcal{D}_\xi^2(W)\left[\frac{\partial\mathcal{L}}{\partial u_{\xi\xi\xi}} - \mathcal{D}_\xi\left(\frac{\partial\mathcal{L}}{\partial u_{\xi\xi\xi\xi}}\right)\right] + \mathcal{D}_\xi^3(W)\left[\frac{\partial\mathcal{L}}{\partial u_{\xi\xi\xi\xi}}\right],$$

where

$$\mathcal{L} = v\left(u_\tau + a_1 u u_\xi - a_2 u_{\xi\xi} + a_3 u_{\xi\xi\xi} + a_4 u_{\xi\xi\xi\xi}\right).$$

Now, we discuss about the self-adjointness of Eq. (1.66) by the following theorem.

Theorem 3 *Eq. (1.66) is neither quasi self-adjoint nor weakly self-adjoint; however Eq. (1.66) is nonlinearly self-adjoint for every solution of*

$$\frac{\partial h(\tau,\xi,u)}{\partial \tau} + a_1 u \frac{\partial h(\tau,\xi,u)}{\partial \xi} + a_2 \frac{\partial^2 h(\tau,\xi,u)}{\partial \xi^2} + a_3 \frac{\partial^3 h(\tau,\xi,u)}{\partial \xi^3} - a_4 \frac{\partial^4 h(\tau,\xi,u)}{\partial \xi^4} = 0. \tag{1.168}$$

Proof: By few computations we can show that Eq. (1.66) is neither quasi self-adjoint nor weakly self-adjoint. To demonstrate the nonlinear self-adjointness, setting $v = h(t,x,u)$ in Eq. (1.167) we get

$$\begin{aligned}
&- h_\tau - u_\tau h_u - a_1 u h_\xi - a_1 u u_\xi h_u - a_2 h_{\xi\xi} - a_2 u_\xi^2 h_{uu} - 2a_2 u_\xi h_{\xi u} - a_2 u_{\xi\xi} h_u \\
&- a_3 h_{\xi\xi\xi} - a_3 u_\xi^3 h_{uuu} - a_3 u_{\xi\xi\xi} h_u - 3a_3 u_\xi^2 h_{\xi uu} - 3a_3 u_\xi h_{\xi\xi u} - 3a_3 u_{\xi\xi} h_{\xi u} \\
&- 3a_3 u_\xi u_{\xi\xi} h_{uu} + a_4 h_{\xi\xi\xi\xi} + a_4 u_{\xi\xi\xi\xi} h_u + 6a_4 u_\xi^2 h_{\xi\xi uu} + a_4 u_\xi^4 h_{uuuu} \\
&+ 4a_4 u_\xi^3 h_{\xi uuu} + 3a_4 u_{\xi\xi}^2 h_{uu} + 4a_4 u_{\xi\xi\xi} h_{\xi u} + 6a_4 u_{\xi\xi} h_{\xi\xi u} + 4a_4 u_\xi h_{\xi\xi\xi u} \\
&+ 6a_4 u_\xi^2 u_{\xi\xi} h_{uuu} + 12a_4 u_\xi u_{\xi\xi} h_{\xi uu} + 4a_4 u_\xi u_{\xi\xi\xi} h_{uu} = 0,
\end{aligned}$$

which yields:

$$\begin{aligned}
&F^* - \lambda\left(u_\tau + a_1 u u_\xi - a_2 u_{\xi\xi} + a_3 u_{\xi\xi\xi} + a_4 u_{\xi\xi\xi\xi}\right) = -\lambda u_\tau - \lambda a_1 u u_\xi + \lambda a_2 u_{\xi\xi} \\
&- \lambda a_3 u_{\xi\xi\xi} - \lambda a_4 u_{\xi\xi\xi\xi} - h_\tau - u_\tau h_u - a_1 u h_\xi - a_1 u u_\xi h_u - a_2 h_{\xi\xi} - a_2 u_\xi^2 h_{uu} \\
&- 2a_2 u_\xi h_{\xi u} - a_2 u_{\xi\xi} h_u - a_3 h_{\xi\xi\xi} - a_3 u_\xi^3 h_{uuu} - a_3 u_{\xi\xi\xi} h_u - 3a_3 u_\xi^2 h_{\xi uu} \\
&- 3a_3 u_\xi h_{\xi\xi u} - 3a_3 u_{\xi\xi} h_{\xi u} - 3a_3 u_\xi u_{\xi\xi} h_{uu} + a_4 h_{\xi\xi\xi\xi} + a_4 u_{\xi\xi\xi\xi} h_u + 6a_4 u_\xi^2 h_{\xi\xi uu} \\
&+ a_4 u_\xi^4 h_{uuuu} + 4a_4 u_\xi^3 h_{\xi uuu} + 3a_4 u_{\xi\xi}^2 h_{uu} + 4a_4 u_{\xi\xi\xi} h_{\xi u} + 6a_4 u_{\xi\xi} h_{\xi\xi u} \\
&+ 4a_4 u_\xi h_{\xi\xi\xi u} + 6a_4 u_\xi^2 u_{\xi\xi} h_{uuu} + 12a_4 u_\xi u_{\xi\xi} h_{\xi uu} + 4a_4 u_\xi u_{\xi\xi\xi} h_{uu}.
\end{aligned} \tag{1.169}$$

Comparing the coefficients for the different derivatives of u we obtain some conditions of which one of them is $\lambda + h_u = 0$. Thus, by setting $\lambda = -h_u$ in (1.169) we get

$$\begin{aligned}
&- h_\tau - a_1 u h_\xi - a_2 h_{\xi\xi} - a_2 u_\xi^2 h_{uu} - 2a_2 u_\xi h_{\xi u} - 2a_2 u_{\xi\xi} h_u - a_3 h_{\xi\xi\xi} \\
&- a_3 u_\xi^3 h_{uuu} - 3a_3 u_\xi^2 h_{\xi uu} - 3a_3 u_\xi h_{\xi\xi u} - 3a_3 u_{\xi\xi} h_{\xi u} - 3a_3 u_\xi u_{\xi\xi} h_{uu} \\
&+ a_4 h_{\xi\xi\xi\xi} + 2a_4 u_{\xi\xi\xi\xi} h_u + 6a_4 u_\xi^2 h_{\xi\xi uu} + a_4 u_\xi^4 h_{uuuu} + 4a_4 u_\xi^3 h_{\xi uuu} \\
&+ 3a_4 u_{\xi\xi}^2 h_{uu} + 4a_4 u_{\xi\xi\xi} h_{\xi u} + 6a_4 u_{\xi\xi} h_{\xi\xi u} + 4a_4 u_\xi h_{\xi\xi\xi u} + 6a_4 u_\xi^2 u_{\xi\xi} h_{uuu} \\
&+ 12a_4 u_\xi u_{\xi\xi} h_{\xi uu} + 4a_4 u_\xi u_{\xi\xi\xi} h_{uu} = 0.
\end{aligned} \tag{1.170}$$

As previously, comparing the coefficients for the different derivatives of u, we have the following condition:

$$h_\tau + a_1 u h_\xi + a_2 h_{\xi\xi} + a_3 h_{\xi\xi\xi} - a_4 h_{\xi\xi\xi\xi} = 0, \tag{1.171}$$

which by solving this system completes the proof.

Here the infinite dimensional Lie algebras of Eq. (1.66) are used to construct the corresponding conservation laws.

Eq. (1.66) admits three-dimensional Lie algebras for nonzero coefficients. Thus, we consider the following three cases:

(i) We first consider the Lie point symmetry generator $X_1 = \frac{\partial}{\partial\tau}$. Components of the conserved vector are given by

$$\begin{aligned}T_1^{\tau} =&a_1 v u u_{\xi} - a_2 v u_{\xi\xi} + a_3 v u_{\xi\xi\xi} + a_4 v u_{\xi\xi\xi\xi},\\ T_1^{\xi} =&- a_1 v u u_{\tau} - a_2 v_{\xi} u_{\tau} + a_2 v u_{\tau\xi} - a_3 u_{\tau} v_{\xi\xi} - a_3 v_{\xi} u_{\tau\xi} - a_3 v u_{\tau\xi\xi}\\ &+ a_4 v_{\xi\xi\xi} u_{\tau} - a_4 v_{\xi\xi} u_{\tau\xi} + a_4 v_{\xi} u_{\tau\xi\xi} - a_4 v u_{\tau\xi\xi\xi}.\end{aligned}$$

Let

$$\begin{aligned}T_1^{\tau}|_{v=1} =&a_1 u u_{\xi} - a_2 u_{\xi\xi} + a_3 u_{\xi\xi\xi} + a_4 u_{\xi\xi\xi\xi}\\ =&\mathcal{D}_{\xi}\left(\frac{1}{2}a_1 u^2 - a_2 u_{\xi} + a_3 u_{\xi\xi} + a_4 u_{\xi\xi\xi}\right),\\ T_1^{\xi}|_{v=1} =&- a_1 u u_{\tau} + a_2 u_{\tau\xi} - a_3 u_{\tau\xi\xi} - a_4 u_{\tau\xi\xi\xi}\\ =&-\mathcal{D}_{\tau}\left(\frac{1}{2}a_1 u^2 - a_2 u_{\xi} + a_3 u_{\xi\xi} + a_4 u_{\xi\xi\xi}\right).\end{aligned}$$

Then transferring the terms $\mathcal{D}_{\xi}(\cdots)$ from T_1^{τ} to T_1^{ξ} provides the null divergence $T_1 = (T_1^{\tau}, T_1^{\xi}) = (0,0)$.

(ii) Using the Lie point symmetry generator $X_2 = \frac{\partial}{\partial\xi}$, the components of the conserved vector are given by

$$\begin{aligned}T_2^{\tau} =&- v u_{\xi},\\ T_2^{\xi} =&v u_{\tau} - a_2 v_{\xi} u_{\xi} - a_3 v_{\xi\xi} u_{\xi} + a_3 v_{\xi} u_{\xi\xi} + a_4 v_{\xi\xi\xi} u_{\xi} - a_4 v_{\xi\xi} u_{\xi\xi} + a_4 v_{\xi} u_{\xi\xi\xi}.\end{aligned}$$

Setting $v = h(\tau,\xi,u) = 1$ into T_2^{τ}, T_2^{ξ} and after reckoning, we have

$$\begin{aligned}T_2^{\tau}|_{v=1} =&- u_{\xi} = \mathcal{D}_{\xi}\left(-u\right),\\ T_2^{\xi}|_{v=1} =&u_{\tau} = -\mathcal{D}_{\tau}\left(-u\right).\end{aligned}$$

Therefore $T_2 = (T_2^{\tau}, T_2^{\xi}) = (0,0)$.

(iii) Using the Lie point symmetry generator $X_3 = a_1\tau\frac{\partial}{\partial\xi} + \frac{\partial}{\partial u}$, one can obtain the conserved vector whose components are

$$\begin{aligned}T_3^{\tau} =&v - a_1 \tau v u_{\xi},\\ T_3^{\xi} =&- a_1 a_2 \tau u_{\xi} v_{\xi} - a_1 a_3 \tau v_{\xi\xi} u_{\xi} + a_1 a_3 \tau v_{\xi} u_{\xi\xi} + a_1 a_4 \tau u_{\xi} v_{\xi\xi\xi}\\ &- a_1 a_4 \tau v_{\xi\xi} u_{\xi\xi} + a_1 a_4 \tau v_{\xi} u_{\xi\xi\xi} + a_1 \tau v u_{\tau} + a_1 u v + a_2 v_{\xi} + a_3 v_{\xi\xi} - a_4 v_{\xi\xi\xi}.\end{aligned}$$

Setting $v = 1$ concludes the previous conserved vectors:

$$\begin{aligned} T_3^{\tau}|_{v=1} =& 1 - a_1 \tau u_{\xi} = \mathcal{D}_{\xi}\left(\xi - a_1 \tau u\right), \\ T_3^{\xi}|_{v=1} =& a_1 u + a_1 \tau u_{\tau} = -\mathcal{D}_{\tau}\left(\xi - a_1 \tau u\right). \end{aligned}$$

Therefore $T_3 = (T_3^{\tau}, T_3^{\xi}) = (0, 0)$.

1.3.3 Conservation laws of the Fornberg-Whitham equation

For the Fornberg-Whitham equation (1.6) we have

$$\mathcal{L} = v\big[u_t - \frac{1}{3}(u_{txx} + u_{xtx} + u_{xxt}) + u_x + uu_x - 3u_x u_{xx} - uu_{xxx}\big].$$

The adjoint equation for Eq. (1.6) is as follows:

$$F^* = \frac{\delta(\mathcal{L})}{\delta u} = \frac{\delta\big(v\big[u_t - \frac{1}{3}(u_{txx} + u_{xtx} + u_{xxt}) + u_x + uu_x - 3u_x u_{xx} - uu_{xxx}\big]\big)}{\delta u},$$

which by some simplifications we get

$$F^* = -v_t - v_x - uv_x + uv_{xxx} + v_{txx} = 0. \tag{1.172}$$

Now, we discuss about the self-adjointness of Eq. (1.6) by the following theorem.

Theorem 4 *Eq .(1.6) is neither quasi self-adjoint nor weak self-adjoint; however Eq. (1.6) is nonlinearly self-adjoint with $h = \chi \in \mathbb{R}$.*

Proof: A straightforward computation shows that Eq. (1.6) is neither quasi self-adjoint nor weak self-adjoint. In order to demonstrate the nonlinear self-adjointness, by setting $v = h(t, x, u)$ in Eq. (1.172) we conclude

$$\begin{aligned} &- u_t h_u - u_x h_u + uh_{xxx} + 2u_t u_x h_{xuu} + u_t u_x^2 h_{uuu} + u_t u_{xx} h_{uu} + 3uu_x h_{xxu} \\ &+ 3uu_x^2 h_{xuu} + uu_x^3 h_{uuu} - uh_x - h_t - h_x - uu_x h_u + 3uu_{xx} h_{xu} + uu_{xxx} h_u \\ &+ 3uu_x u_{xx} h_{uu} + h_{txx} + 2u_x h_{txu} + u_t h_{xxu} + 2u_x u_{tx} h_{uu} + h_u u_{txx} \\ &+ 2u_{tx} h_{xu} + h_{tuu} u_x^2 + h_{tu} u_{xx} = 0, \end{aligned}$$

which yields:

$$\begin{aligned} &F^* - \lambda\left(u_t - u_{xxt} + u_x + uu_x - 3u_x u_{xx} - uu_{xxx}\right) = -u_t h_u - u_x h_u + uh_{xxx} \\ &- \lambda u_t + \lambda u_{txx} - \lambda u_x + 2u_t u_x h_{xuu} + u_t u_x^2 h_{uuu} + u_t u_{xx} h_{uu} - \lambda uu_x \\ &+ \lambda uu_{xxx} + 3\lambda u_x u_{xx} + 3uu_x h_{xxu} + 3uu_x^2 h_{xuu} + uu_x^3 h_{uuu} - uh_x - h_t \\ &- h_x - uu_x h_u + 3uu_{xx} h_{xu} + uu_{xxx} h_u + 3uu_x u_{xx} h_{uu} + h_{txx} + 2u_x h_{txu} \\ &+ u_t h_{xxu} + 2u_x u_{tx} h_{uu} + h_u u_{txx} + 2u_{tx} h_{xu} + u_x^2 h_{tuu} + h_{tu} u_{xx} = 0. \end{aligned} \tag{1.173}$$

Comparing the coefficients for the different derivatives of u we obtain some conditions of which one of them is $\lambda + h_u = 0$. Thus, by setting $\lambda = -h_u$ in (1.173) we get

$$\begin{aligned}
&h_{tu}u_{xx} + 2h_{txu}u_x + h_{tuu}u_x^2 + u_t h_{xxu} + 2h_{xu}u_{tx} - h_x u + h_{xxx}u - 3h_u u_x u_{xx} \\
&+ 2h_{xuu}u_t u_x + h_{uuu}u_t u_x^2 + h_{uu}u_t u_{xx} + 2h_{uu}u_x u_{tx} + 3h_{xxu}uu_x + 3h_{xuu}uu_x^2 \\
&+ 3h_{xu}uu_{xx} + h_{uuu}uu_x^3 - h_t - h_x + h_{txx} + 3h_{uu}uu_x u_{xx} = 0. \qquad (1.174)
\end{aligned}$$

From the coefficient of $u_x u_{xx}$ we have $h_u = 0$, which concludes that the function h is independent of u, in other words $h = h(t, x)$. Also, comparing the coefficients for the different derivatives of u in Eq. (1.174), we have the following conditions:

$$h_{xxx} - h_x = 0, \quad -h_t - h_x + h_{txx} = 0,$$

which by solving them we have $h = \chi \in \mathbb{R}$.□

Setting $t = x^1$ and $x = x^2$, the conservation law will be written

$$\mathcal{D}_t(T_i^t) + \mathcal{D}_x(T_i^x) = 0, \quad i = 1, 2, 3.$$

We recall that Eq. (1.6) admits a three-dimensional Lie algebra; thus we consider the following three cases:

(i) We first consider the Lie point symmetry generator $X_1 = \frac{\partial}{\partial t}$. The components of the conserved vector are given by

$$\begin{aligned}
T_1^t =& -\frac{2}{3}vu_{txx} + vu_x + vuu_x - 3vu_x u_{xx} - vuu_{xxx} + \frac{1}{3}(u_t v_{xx} - u_{tx}v_x), \\
T_1^x =& -vu_t - vuu_t + vu_t u_{xx} - v_x u_t u_x + v_{xx}uu_t + \frac{2}{3}(u_t v_{tx} + vu_{ttx}) \\
& - \frac{1}{3}(v_x u_{tt} + v_t u_{tx}) + 2vu_x u_{tx} - v_x uu_{tx} + vuu_{txx}.
\end{aligned}$$

Therefore

$$\begin{aligned}
T_1^t|_{v=1} =& \mathcal{D}_x\left(\frac{1}{2}(u^2 - u_x^2) - \frac{2}{3}u_{tx} + u - uu_{xx}\right), \\
T_1^x|_{v=1} =& -\mathcal{D}_t\left(\frac{1}{2}(u^2 - u_x^2) - \frac{2}{3}u_{tx} + u - uu_{xx}\right).
\end{aligned}$$

Then transferring the terms $\mathcal{D}_x(\cdots)$ from T_1^t to T_1^x provides the null divergence $T_1 = (T_1^t, T_1^x) = (0, 0)$.

(ii) Using the Lie point symmetry generator $X_2 = \frac{\partial}{\partial x}$, the components of the conserved vector are given by

$$\begin{aligned}
T_2^t =& -vu_x + \frac{1}{3}(u_x v_{xx} - v_x u_{xx} + vu_{xxx}), \\
T_2^x =& vu_t - \frac{1}{3}(vu_{txx} + v_x u_{tx} + v_t u_{xx}) - v_x u_x^2 + v_{xx}uu_x - v_x uu_{xx} + \frac{2}{3}v_{tx}u_x.
\end{aligned} \qquad (1.175)$$

Setting $v=1$ into (1.175) and after reckoning, we have

$$\begin{aligned} T_2^t|_{v=1} =& \mathcal{D}_x\left(-u+\frac{1}{3}u_{xx}\right), \\ T_2^x|_{v=1} =& -\mathcal{D}_t\left(-u+\frac{1}{3}u_{xx}\right), \end{aligned}$$

from where we can obtain the conserved current $T_2=(T_2^t,T_2^x)=(0,0)$.

(iii) Using the Lie point symmetry generator $X_3=t\frac{\partial}{\partial x}+\frac{\partial}{\partial u}$, one can obtain the conserved vector whose components are

$$\begin{aligned} T_3^t =& v-tu_xv-\frac{1}{3}(v_{xx}-tv_{xx}u_x+tv_xu_{xx}-tvu_{xxx}), \\ T_3^x =& v+vu-uv_{xx}+tuu_xv_{xx}-tuv_xu_{xx}-tv_xu_x^2+tvu_t \\ & -\frac{1}{3}(vu_{xx}+tvu_{txx}+tv_xu_{tx}+tv_tu_{xx})+\frac{2}{3}(v_xu_x-v_{tx}+tu_xv_{tx}). \end{aligned}$$

Substituting $v=1$ into the components above, we obtain

$$\begin{aligned} T_3^t|_{v=1} =& \mathcal{D}_x\left(x-tu+\frac{1}{3}tu_{xx}\right), \\ T_3^x|_{v=1} =& 1-\mathcal{D}_t\left(x-tu+\frac{1}{3}tu_{xx}\right), \end{aligned}$$

which provides the nontrivial components $T_3^t=0$ and $T_3^x=1$.

1.3.4 Conservation laws of the mKdV-KP equation

Here, we try to obtain the conservation laws of the mKdV-KP equation. The adjoint equation for Eq. (1.92) is as follows:

$$\begin{aligned} & F^*(t,x,u,v,u_{(1)},v_{(1)},u_{(2)},v_{(2)},u_{(3)},v_{(3)},u_{(4)},v_{(4)}) = \frac{\delta(vF)}{\delta u} \\ & = \frac{\delta\left(v\left[\left(u_t-\frac{3}{2}u_x+6u^2u_x+u_{xxx}\right)_x+u_{yy}\right]\right)}{\delta u}, \end{aligned}$$

from which by some simplifications we can obtain

$$F^*=v_{tx}-\frac{3}{2}v_{xx}+6u^2v_{xx}+v_{xxxx}+v_{yy}=0.$$

Here we recall that Eq. (1.92) admits five-dimensional Lie algebras; thus we consider the following five cases.

(i) We first consider the vector field $V_1 = \frac{\partial}{\partial t}$ for Eq. (1.92). The conserved vector is given by

$$\begin{aligned}
T_1^t =& -\frac{3}{2}vu_{xx} + 12vuu_x^2 + 6vu^2u_{xx} + vu_{xxxx} + vu_{yy} + u_tv_x,\\
T_1^x =& -12vuu_tu_x - \frac{3}{2}u_tv_x + 6u^2u_tv_x + u_tv_t + u_tv_{xxx} - vu_{tt}\\
&+ \frac{3}{2}vu_{tx} - 6vu^2u_{tx} + u_{tx}v_{xx} + u_{txx}v_x - vu_{txxx},\\
T_1^y =& u_tv_y - vu_{ty}.
\end{aligned}$$

(ii) Using the Lie point symmetry generator $V_2 = \frac{\partial}{\partial x}$, the components of the conserved vector are given by

$$\begin{aligned}
T_2^t =& u_xv_x - vu_{xx},\\
T_2^x =& vu_{yy} - \frac{3}{2}u_xv_x + 6u^2u_xv_x + u_xv_t + u_xv_{xxx} + u_{xx}v_{xx} + v_xu_{xxx},\\
T_2^y =& u_xv_y - vu_{xy}.
\end{aligned}$$

(iii) Using $V_3 = \frac{\partial}{\partial y}$, one can obtain the conserved vector with the following components:

$$\begin{aligned}
T_3^t =& u_yv_x - vu_{xy},\\
T_3^x =& -12uvu_xu_y - \frac{3}{2}u_yv_x + 6u^2u_yv_x + u_yv_t + u_yv_{xxx} - vu_{ty}\\
&+ \frac{3}{2}vu_{xy} - 6vu^2u_{xy} + u_{xy}v_{xx} + v_xu_{xxy} - vu_{xxxy},\\
T_3^y =& vu_{tx} - \frac{3}{2}vu_{xx} + 12uvu_x^2 + 6vu^2u_{xx} + vu_{xxxx} + u_yv_y.
\end{aligned}$$

(iv) The Lie point symmetry generator $V_4 = y\frac{\partial}{\partial x} - 2t\frac{\partial}{\partial y}$ yields the conserved vector with components:

$$\begin{aligned}
T_4^t =& yu_xv_x - 2tv_xu_y - yvu_{xx} + 2tvu_{xy},\\
T_4^x =& 24tuvu_xu_y + 2vu_y + yvu_{yy} + yu_xv_t + yu_xv_{xxx} - 2tu_yv_t - 2tu_yv_{xxx}\\
&+ 2tvu_{ty} + yu_{xx}v_{xx} - 2tu_{xy}v_{xx} - 2tv_xu_{xxy} + yv_xu_{xxx} + 2tvu_{xxxy}\\
&+ 3tu_yv_x - \frac{3}{2}yu_xv_x - 3tvu_{xy} + 6yu^2u_xv_x - 12tu^2u_yv_x + 12tvu^2u_{xy},\\
T_4^y =& -2tvu_{tx} + 3tvu_{xx} - 24tvuu_x^2 - 12tvu^2u_{xx}\\
&- 2tvu_{xxxx} + yu_xv_y - 2tv_yu_y - yvu_{xy} - vu_x.
\end{aligned}$$

(v) Finally, making the same procedures to $V_5 = -3t\frac{\partial}{\partial t} + (3t - x)\frac{\partial}{\partial x} - 2y\frac{\partial}{\partial y} + u\frac{\partial}{\partial u}$, we obtain the following conserved quantities

$$\begin{aligned}
T_5^t =& \frac{3}{2}tvu_{xx} - 36tuvu_x^2 - 18tvu^2u_{xx} - 3tvu_{xxxx} - 3tvu_{yy} - uv_x - 3tu_tv_x \\
&+ 3tu_xv_x - xu_xv_x - 2yu_yv_x + xvu_{xx} + 2yvu_{xy} + 2vu_x, \\
T_5^x =& - 6vu_x + \frac{3}{2}uv_x + 36tuvu_tu_x + 24yuvu_xu_y - 6u^3v_x - uv_t - uv_{xxx} \\
&+ 4vu_t - 2u_xv_{xx} - 3v_xu_{xx} + 4vu_{xxx} - xvu_{yy} + 24vu^2u_x - 3tu_tv_t \\
&+ 3tu_xv_t + 3tu_xv_{xxx} - xu_xv_t - xu_xv_{xxx} - 2yu_yv_t - 2yu_yv_{xxx} \\
&+ 3tvu_{tt} + 2yvu_{ty} + 3tu_{xx}v_{xx} - xu_{xx}v_{xx} - 3tu_{tx}v_{xx} - 2yu_{xy}v_{xx} \\
&- 3tv_xu_{txx} - 2yv_xu_{xxy} + 3tv_xu_{xxx} - xv_xu_{xxx} + 3tvu_{txxx} + 2yvu_{xxxy} \\
&- \frac{9}{2}tvu_{tx} - 3yvu_{xy} + \frac{9}{2}tu_tv_x - \frac{9}{2}tu_xv_x + \frac{3}{2}xu_xv_x + 3yu_yv_x + 3tvu_{yy} \\
&- 18tu^2u_tv_x + 18tu^2u_xv_x - 6xu^2u_xv_x - 12yu^2u_yv_x + 18tvu^2u_{tx} \\
&+ 12yvu^2u_{xy}, \\
T_5^y =& - 2yvu_{tx} + 3yvu_{xx} - 24yuvu_x^2 - 12yu^2vu_{xx} - 2yvu_{xxxx} - uv_y - 3tu_tv_y \\
&+ 3tu_xv_y - xu_xv_y - 2yu_yv_y - 3tvu_{xy} + xvu_{xy} + 3tvu_{ty} + 3vu_y.
\end{aligned}$$

Chapter 2

Group analysis and exact solutions of fractional partial differential equations

The basic idea of the Lie symmetry analysis is the consideration of the tangent structural equations under one or several parameter transformation groups in conjunction with the system of differential equations. It is appropriate to mention here that for nonlinear partial differential equations (PDEs) with two independent variables exhibiting solitons, the Lie symmetry analysis not only helps to study their group theoretical properties but also to derive several mathematical characteristics related with their complete integrability [117, 52, 133].

In recent years, the study of fractional ordinary differential equations (FODEs) and fractional partial differential equations (FPDEs) has attracted much attention due to an exact description of nonlinear phenomena in fluid mechanics, viscoelasticity, biology, physics, engineering and other areas of science. In reality, a physical phenomenon may depend not only on the time instant but also on the previous time history, which can be successfully modeled by using the theory of derivatives and integrals of fractional order. The time and space FPDEs are obtained by replacing the integer order time and space derivatives in PDEs by the fractional derivative of order $\alpha > 0$.

2.1 Basic theory of fractional differential equations

This section deals with the preliminaries about the problems arising in the area of fractional calculus – a branch of mathematics that is, in a certain sense, as old as classical calculus as we know it today.

The basic idea behind fractional calculus is intimately related to a classical standard result from (classical) differential and integral calculus, the fundamental theorem [163].

Theorem 5 ***(Fundamental Theorem of Classical Calculus)*** *Let* $f \in C[a,b]$, $Df(x) := f'(x)$, ${}_aJ_x f(x) := \int_a^x f(s)ds$ *and* $F : [a,b] \to \mathbb{R}$ *be defined by*

$$F(x) := {}_aJ_x f(x).$$

Then, F *is differentiable and*

$$DF = f.$$

Therefore, we have a very close relation between differential operators and integral operators. It is one of the goals of fractional calculus to retain this relation in a suitably generalized sense. Hence there is also a need to deal with fractional integral operators, and actually it turns out to be useful to discuss these first before coming to fractional differential operators.

For $n \in \mathbb{N}$ we use the symbols D^n and ${}_aJ_x^n$ to denote the n-fold iterates of D and ${}_aJ_x$, respectively, i.e., we set $D^1 := D$, ${}_aJ_x^1 := {}_aJ_x$, and $D^n := DD^{n-1}$ and ${}_aJ_x^n := {}_aJ_x \, {}_aJ_x^{n-1}$ for $n \geq 2$.

Following the outline given above, we begin with the integral operator J_a^n. In the case $n \in \mathbb{N}$, it is well known (and easily proved by induction) [107, 185] that we can replace the recursive definition of integral operator by the following explicit formula.

Theorem 6 *Let* f *be Riemann integrable on* $[a,b]$. *Then, for* $a \leq x \leq b$ *and* $n \in \mathbb{N}$, *we have*

$${}_aJ_x^n f(x) = \frac{1}{(n-1)!} \int_a^x (x-s)^{n-1} f(s)ds.$$

In this book, we also frequently use Euler's gamma function $\Gamma : (0,\infty) \to \mathbb{R}$, defined by

$$\Gamma(x) := \int_0^\infty t^{x-1} e^{-t} dt,$$

which has a useful property

$$\Gamma(n+1) = n!,$$

for $n \in \mathbb{N}$, and

$$\Gamma(n+1) = n\Gamma(n),$$

for $n \in \mathbb{R}$. Moreover, like the exponential function in the theory of integer-order differential equations, the Mittag-Leffler function plays a very important role in the theory of fractional differential equations.

Definition 12 *[47, 1] Let* $n > 0$. *The function* E_n *defined by*

$$E_n(z) := \sum_{j=0}^{\infty} \frac{z^j}{\Gamma(jn+1)},$$

whenever the series converges, is called the Mittag-Leffler function of order n.

We immediately notice that

$$E_1(z) = \sum_{j=0}^{\infty} \frac{z^j}{\Gamma(j+1)} = \sum_{j=0}^{\infty} \frac{z^j}{j!} = exp(z).$$

In view of the above considerations, the following concept seems rather natural.

Definition 13 *[47, 109, 156] Let $n \in \mathbb{R}_+$. The operator ${}_aJ_x^n$, defined on $L_1[a,b]$ by*

$${}_aJ_x^n f(x) := \frac{1}{\Gamma(n)} \int_a^x (x-s)^{n-1} f(s) ds,$$

for $a \le x \le b$, is called the Riemann-Liouville fractional integral operator of order n.

For $n = 0$, we set ${}_aJ_x^0 := I$, the identity operator.

In the special case when $f(x) = (x-a)^\beta$, for some $\beta > -1$ and $n > 0$ we have

$${}_aJ_x^n f(x) = \frac{\Gamma(\beta+1)}{\Gamma(n+\beta+1)} (x-a)^{n+\beta}.$$

In view of the well known corresponding result in the case $n \in \mathbb{N}$, this result is precisely what one would expect from a sensible generalization of the integral operator.

Theorem 7 *[47] Let f be analytic in $(a-h, a+h)$ for some $h > 0$, and let $n > 0$. Then*

$${}_aJ_x^n f(x) = \sum_{k=0}^{\infty} \frac{(-1)^k (x-a)^{k+n}}{k!(n+k)\Gamma(n)} D^k f(x),$$

for $x \in [a, a+\frac{h}{2})$, and

$${}_aJ_x^n f(x) = \sum_{k=0}^{\infty} \frac{(x-a)^{k+n}}{\Gamma(k+n+1)} D^k f(a),$$

for $x \in [a, a+h)$.

Having established these fundamental properties of Riemann-Liouville integral operators, we now come to the corresponding differential operators.

Definition 14 *[47] Let $n \in \mathbb{R}_+$ and $m = \lceil n \rceil$. The operator ${}_a^{RL}D_x^n$, defined by*

$${}_a^{RL}D_x^n f := D^m \, {}_aJ_x^{m-n} f,$$

is called the Riemann-Liouville fractional differential operator of order n.

For $n = 0$, we set ${}_a^{RL}D_x^0 := I$, the identity operator.

Lemma 2.1.1 *[47] Let $n \in \mathbb{R}_+$ and let $m \in \mathbb{N}$ such that $m > n$. Then,*

$$ {}_a^{RL}D_x^n = D^m \; {}_aJ_x^{m-n}. $$

Proof 1 *Our assumption on m concludes that $m \geq n$. Thus,*

$$ D^m \; {}_aJ_x^{m-n} = D^{\lceil n \rceil} D^{m-\lceil n \rceil} \; {}_aJ_x^{m-\lceil n \rceil} \; {}_aJ_x^{\lceil n \rceil - n} = D^{\lceil n \rceil} \; {}_aJ_x^{\lceil n \rceil - n} = {}_a^{RL}D_x^n, $$

in view of the semigroup property of fractional integration.

Therefore, when $f(x) = (x-a)^\beta$, for some $\beta > -1$ and $n > 0$ we have

$$ {}_a^{RL}D_x^n f(x) = D^{\lceil n \rceil} \; {}_aJ_x^{\lceil n \rceil - n} f(x) = \frac{\Gamma(\beta+1)}{\Gamma(\lceil n \rceil - n + \beta + 1)} D^{\lceil n \rceil} (x-a)^{\lceil n \rceil - n + \beta}. $$

Specifically, if $n - \beta \in \mathbb{N}$ then $\lceil n \rceil - (n - \beta) \in \{0, 1, \ldots, \lceil n \rceil - 1\}$, and so

$$ D^{\lceil n \rceil} (x-a)^{\lceil n \rceil - n + \beta} = 0. $$

On the other hand, if $n - \beta \notin \mathbb{N}$, we obtain

$$ {}_a^{RL}D_x^n (x-a)^\beta = \frac{\Gamma(\beta+1)}{\Gamma(\beta+1-n)} (x-a)^{\beta-n}. $$

Moreover, for $n > 0$ we have

$$ {}_a^{RL}D_x^n C = \frac{C}{\Gamma(1-n)} (x-a)^{-n}. $$

Theorem 8 *[47] Let $n \geq 0$. Then, for every $f \in L_1[a,b]$,*

$$ {}_a^{RL}D_x^n \; {}_aJ_x^n f = f, $$

almost everywhere.

Theorem 9 *[47] Let f be analytic in $(a-h, a+h)$ for some $h > 0$, and let $n > 0$, $n \notin \mathbb{N}$. Then*

$$ {}_a^{RL}D_x^n f(x) = \sum_{k=0}^{\infty} \binom{n}{k} \frac{(x-a)^{k-n}}{\Gamma(k+1-n)} D^k f(x), $$

for $x \in [a, a + \frac{h}{2})$, and

$$ {}_a^{RL}D_x^n f(x) = \sum_{k=0}^{\infty} \frac{(x-a)^{k-n}}{\Gamma(k+1-n)} D^k f(a), $$

for $x \in [a, a+h)$.

Theorem 10 *[47] Let f_1 and f_2 be two functions defined on $[a,b]$ such that ${}_a^{RL}D_x^n f_1(x)$ and ${}_a^{RL}D_x^n f_2(x)$ exist almost everywhere. Moreover, let $c_1,\ c_2 \in \mathbb{R}$. Then, ${}_a^{RL}D_x^n(c_1 f_1 + c_2 f_2)$ exists almost everywhere, and*

$$ {}_a^{RL}D_x^n(c_1 f_1 + c_2 f_2) = c_1\ {}_a^{RL}D_x^n f_1 + c_2\ {}_a^{RL}D_x^n f_2. $$

When it comes to products of functions, the situation is completely different. In the classical case, if $n \in \mathbb{N}$ and $f, g \in C^n[a,b]$ we have

$$ D^n(fg) = \sum_{k=0}^{n} \binom{n}{k} (D^k f)(D^{n-k} g), $$

which is well known as Leibniz's formula. We point out two special properties of this result: The formula is symmetric, i.e., we may interchange f and g on both sides of the equation without altering the expression, and in order to evaluate the nth derivative of the product fg, we only need derivatives up to the order n of both factors. In particular, none of the factors needs to have an $(n+1)$st derivative. The following theorem transfers Leibniz's formula to the fractional setting, and it is immediately evident that both these properties are lost.

Theorem 11 *[47, 156] Let $n > 0$, and assume that f and g are analytic on $(a-h, a+h)$ with some $h > 0$. Then,*

$$ {}_a^{RL}D_x^n(fg)(x) = \sum_{k=0}^{\lfloor n \rfloor} \binom{n}{k} ({}_a^{RL}D_x^k f)(x)({}_a^{RL}D_x^{n-k} g)(x) $$

$$ + \sum_{k=\lfloor n \rfloor + 1}^{\infty} \binom{n}{k} ({}_a^{RL}D_x^k f)(x)({}_aJ_x^{k-n} g)(x), $$

for $a < x < a + \frac{h}{2}$.

In spite of all these differences we recover the classical result from the fractional result by using an integer value for n because then the binomial coefficients $\binom{n}{k}$ are zero for $k > n$, so that the second sum (the one that causes all the differences) vanishes.

Let, $n \geq 0$ and $m = \lceil n \rceil$. Then, we define the Caputo fractional derivative by [156]

$$ {}_a^{C}D_x^n f(x) = {}_aJ_x^{m-n} D^m f(x), $$

whenever $D^m f \in L_1[a,b]$.

In general, the Caputo and the Riemann-Liouville fractional derivatives do not coincide. The connections between them are given as [156]

$$ {}_a^{C}D_x^n f(x) = {}_a^{RL}D_x^n \left(f(x) - \sum_{k=0}^{\lfloor n \rfloor} \frac{(x-a)^k}{k!} D^k f(a) \right). $$

Note that the Caputo derivative of a constant function is zero.

For the Caputo derivative, if $f(x) = (x-a)^{\beta}$ and $\beta \geq 0$, then

$$ {}_{a}^{C}D_{x}^{n} f(x) = \begin{cases} 0; & \text{if} \quad \beta \in \{0,1,2,\ldots,\lfloor n \rfloor\}, \\ \frac{\Gamma(\beta+1)}{\Gamma(\beta+1-n)}(x-a)^{\beta-n}; & \text{if } \beta \in \mathbb{N} \text{ and } \beta \geq \lceil n \rceil \\ & \text{or } \beta \notin \mathbb{N} \text{ and } \beta \geq \lfloor n \rfloor. \end{cases} $$

Another fractional derivative which we use in this book is the Erdélyi-Kober fractional differential operator defined by

$$ \left(\mathcal{P}_{\beta}^{\tau,\alpha}\mathcal{F}\right)(\zeta) := \prod_{j=0}^{n-1}\left(\tau + j - \frac{1}{\beta}\zeta\frac{d}{d\zeta}\right)\left(\mathcal{K}_{\beta}^{\tau+\alpha,n-\alpha}\mathcal{F}\right)(\zeta), \tag{2.1} $$

where

$$ n = \begin{cases} \lceil \alpha \rceil, & \alpha \notin \mathbb{N} \\ \alpha, & \alpha \in \mathbb{N} \end{cases} \tag{2.2} $$

and

$$ \left(\mathcal{K}_{\beta}^{\tau,\alpha}\mathcal{F}\right)(\zeta) := \begin{cases} \frac{1}{\Gamma(\alpha)}\int_{1}^{\infty}(s-1)^{\alpha-1}s^{-(\tau+\alpha)}\mathcal{F}(\zeta s^{\frac{1}{\beta}})ds, & \alpha > 0, \\ \mathcal{F}(\zeta), \quad \alpha = 0, \end{cases} \tag{2.3} $$

denotes the Erdélyi-Kober fractional integral operator.

2.2 Group analysis of fractional differential equations

In this section, we discuss the Lie symmetry analysis to FPDEs. Consider a FPDE having the form [18, 71, 77]:

$$ \partial_t^{\alpha} u = F(x,t,u,u_x,u_{xx}), \tag{2.4} $$

where $\partial_t^{\alpha} u := {}^{RL}D_t^{\alpha} u$ stands for Riemann-Liouville derivative of order $\alpha \in (0,1)$. One-parameter Lie group of infinitesimal transformations of this equation is:

$$ \begin{cases} \bar{t} = \bar{t}(x,t,u;\epsilon), \\ \bar{x} = \bar{x}(x,t,u;\epsilon), \\ \bar{u} = \bar{u}(x,t,u;\epsilon), \end{cases} \tag{2.5} $$

where ϵ is the group parameter and its associated Lie algebra is spanned by

$$ V = \xi^1(x,t,u)\frac{\partial}{\partial t} + \xi^2(x,t,u)\frac{\partial}{\partial x} + \phi(x,t,u)\frac{\partial}{\partial u}, \tag{2.6} $$

where

$$ \xi^1 = \frac{d\bar{t}}{d\epsilon}|_{\epsilon=0}, \quad \xi^2 = \frac{d\bar{x}}{d\epsilon}|_{\epsilon=0}, \quad \phi = \frac{d\bar{u}}{d\epsilon}|_{\epsilon=0}. $$

If the vector field (2.6) generates a symmetry of (2.4), then V must satisfy the Lie symmetry condition

$$Pr^{(\alpha,2)}V(\Delta)|_{\Delta=0} = 0, \quad \Delta = \partial_t^\alpha u - F.$$

The prolongation operator $Pr^{(\alpha,2)}V$ takes the form

$$Pr^{(\alpha,2)}V = V + \phi_\alpha^0 \frac{\partial}{\partial_t^\alpha u} + \phi^x \frac{\partial}{\partial_{u_x}} + \phi^{xx} \frac{\partial}{\partial_{u_{xx}}},$$

where

$$\begin{aligned}
&\phi^x = D_x(\phi) - u_x D_x(\xi^2) - u_t D_x(\xi^1),\\
&\phi^{xx} = D_x(\phi^x) - u_{xt} D_x(\xi^1) - u_{xx} D_x(\xi^2),\\
&\phi_\alpha^0 = D_t^\alpha(\phi) + \xi^2 D_t^\alpha(u_x) - D_t^\alpha(\xi^2 u_x) + D_t^\alpha(D_t(\xi^1)u)\\
&- D_t^{\alpha+1}(\xi^1 u) + \xi^1 D_t^{\alpha+1}(u).
\end{aligned}$$

The invariance condition

$$\xi^1(x,t,u)|_{t=0} = 0 \tag{2.7}$$

is necessary to the transformations (2.5), because of the conservative property of fractional derivative operator (2.7).
The αth extended infinitesimal has the form:

$$\begin{aligned}
&\phi_\alpha^0 = D_t^\alpha(\phi) + \xi^2 D_t^\alpha(u_x) - D_t^\alpha(\xi^2 u_x) + D_t^\alpha(D_t(\xi^1)u)\\
&- D_t^{\alpha+1}(\xi^1 u) + \xi^1 D_t^{\alpha+1}(u),
\end{aligned} \tag{2.8}$$

where the operator D_t^α expresses the total fractional derivative operator. The generalized Leibniz rule in the fractional sense is given by

$$D_t^\alpha\left[u(t)v(t)\right] = \sum_{n=0}^{\infty} \binom{a}{n} D_t^{\alpha-n}u(t) D_t^n v(t), \quad \alpha > 0, \tag{2.9}$$

where

$$\binom{a}{n} = \frac{(-1)^{n-1}\alpha\Gamma(n-\alpha)}{\Gamma(1-\alpha)\Gamma(n+1)}. \tag{2.10}$$

Thus from (2.9) we can rewrite (2.8) as follows:

$$\begin{aligned}
\phi_\alpha^0 =& D_t^\alpha(\phi) - \alpha D_t(\xi^1)\frac{\partial^\alpha u}{\partial t^\alpha} - \sum_{n=1}^{\infty}\binom{a}{n} D_t^n(\xi^2) D_t^{\alpha-n}(u_x)\\
&- \sum_{n=1}^{\infty}\binom{a}{n+1} D_t^{n+1}(\xi^1) D_t^{\alpha-n}(u).
\end{aligned} \tag{2.11}$$

Also from the chain rule we have

$$\frac{d^m f(g(t))}{dt^m} = \sum_{k=0}^{m}\sum_{r=0}^{k}\binom{k}{r}\frac{1}{k!}[-g(t)]^r \frac{d^m}{dt^m}[g(t)^{k-r}]\frac{d^k f(g)}{dg^k}, \tag{2.12}$$

and setting $f(t) = 1$, one can get

$$D_t^\alpha(\phi) = \frac{\partial^\alpha \phi}{\partial t^\alpha} + \phi_u \frac{\partial^\alpha u}{\partial t^\alpha} - u\frac{\partial^\alpha \phi_u}{\partial t^\alpha} + \sum_{n=1}^{\infty} \binom{a}{n} \frac{\partial^n \phi_u}{\partial t^n} D_t^{\alpha-n}(u) + \mu, \quad (2.13)$$

where

$$\mu = \sum_{n=2}^{\infty}\sum_{m=2}^{n}\sum_{k=2}^{m}\sum_{r=0}^{k-1} \binom{a}{n}\binom{n}{m}\binom{k}{r}\frac{1}{k!} \times \frac{t^{n-\alpha}}{\Gamma(n+1-\alpha)}[-u]^r \frac{\partial^m}{\partial t^m}[u^{k-r}]\frac{\partial^{n-m+k}\phi}{\partial t^{n-m}\partial u^k}. \quad (2.14)$$

Therefore

$$\begin{aligned}\phi_\alpha^0 = &\frac{\partial^\alpha \phi}{\partial t^\alpha} + (\phi_u - \alpha D_t(\xi^1))\frac{\partial^\alpha u}{\partial t^\alpha} - u\frac{\partial^\alpha \phi_u}{\partial t^\alpha} + \mu \\ &+ \sum_{n=1}^{\infty}[\binom{a}{n}\frac{\partial^\alpha \phi_u}{\partial t^\alpha} - \binom{a}{n+1}D_t^{n+1}(\xi^1)]D_t^{\alpha-n}(u) \\ &- \sum_{n=1}^{\infty}\binom{a}{n}D_t^n(\xi^2)D_t^{\alpha-n}(u_x).\end{aligned}$$

2.3 Group analysis of time-fractional Fokker-Planck equation

One of the widely used equations of statistical physics is the Fokker-Planck (FP) equation (named after Adriaan Fokker and Max Planck) which describes the time evolution of the probability density function of position and velocity of a particle.

The general fractional FP equation for the motion of a concentration field $u(x,t)$ of one space variable x at time t has the form

$$\partial_t^\alpha u(x,t) = [\frac{\partial}{\partial x}\mathcal{V}_1(x) + \frac{\partial^2}{\partial x^2}\mathcal{V}_2(x)]u(x,t), \quad (2.15)$$

where $\mathcal{V}_1(x) > 0$ is the diffusion coefficient and $\mathcal{V}_2(x) > 0$ is the drift coefficient. Note that the drift and diffusion coefficients may also depend on time. There is a more general form of FP equation which is called nonlinear FP equation [54, 55] and has important applications in various areas such as neuroscience, plasma physics, surface physics, population dynamic, biophysics, neuroscience, nonlinear hydrodynamics, polymer physics, laser physics, pattern formation, psychology, engineering and marketing. A more general case of Eq. (2.15) has the form

$$\partial_t^\alpha u(x,t) = [\frac{\partial}{\partial x}\mathcal{V}_1(x,t,u) + \frac{\partial^2}{\partial x^2}\mathcal{V}_2(x,t,u)]u(x,t), \quad (2.16)$$

which arise in modeling of anomalous diffusive and subdiffusive systems, continuous time random walks, unification of diffusion and wave propagation phenomenon. Eq. (2.16) is solvable using the Lie group analysis, but it should be considered many subcases for different values of $\mathcal{V}_1(x,t,u)$ and $\mathcal{V}_2(x,t,u)$. Thus, in this paper we focus on the Lie group analysis of time-fractional FP equation [71] with cases $\mathcal{V}_1(x,t,u) = u^{\kappa}(x,t)$ and $\mathcal{V}_2(x,t,u) = u^{\nu}(x,t)$, being used to characterize subdiffusion [71]

$$\partial_t^{\alpha} u(x,t) = \big[\frac{\partial}{\partial x} u^{\kappa}(x,t) + \frac{\partial^2}{\partial x^2} u^{\nu}(x,t)\big] u(x,t), \tag{2.17}$$

where $u(x,t)$ is the probability density and $\partial_t^{\alpha} u := {}^{RL}D_t^{\alpha} u$ stands for Riemann-Liouville derivative.

The FP equation with fractional derivatives has been investigated by some authors. Deng in [45] developed the finite element method for the numerical resolution of the space and time-fractional FP equation and then proved that the convergent order is $O(k^{2-\alpha} + h^{\mu})$, where k is the time step size and h the space step size. In [44], first the time-fractional FP equation (Riemann-Liouville derivative) is converted into a time FODE. Then a combination of predictor-corrector approach and method of lines is utilized for numerically solving FODE with the numerical error $O\left(k^{min\{1+2\alpha,2\}}\right) + O\left(h^2\right)$. Chen et al. [36] examined the finite difference approximation and energy method to solve a class of initial-boundary value problems for the fractional FP equation on a finite domain. Odibat and Momani [147] have solved the space- and time-fractional FP equation using the variational iteration method and the Adomian decomposition method with the fractional derivatives described in the Caputo sense. Mousa and Kaltayev [136] have applied the He's homotopy perturbation method for solving fractional FP equations effectively.

According to the Lie theory, applying the prolongation $Pr^{(\alpha,2)}V$ to Eq. (2.17), we can get the following invariance criterion:

$$\begin{aligned} &\phi_{\alpha}^{0} - (\kappa+1)u^{\kappa}\phi^{x} - (\kappa+1)\kappa u^{\kappa-1}\phi u_x - (\nu^2+\nu)(\nu-1)\phi u^{\nu-2}u_x^2 \\ &- 2(\nu^2+\nu)u^{\nu-1}u_x\phi^{x} - \nu(\nu+1)u^{\nu-1}\phi u_{xx} - (\nu+1)u^{\nu}\phi^{xx} = 0. \end{aligned} \tag{2.18}$$

Therefore, we obtain

$$\xi^1 = (\nu - 2\kappa)tc_2, \quad \xi^2 = c_1 + \alpha(\nu-\kappa)xc_2, \quad \phi = \alpha u c_2,$$

where c_1 and c_2 are arbitrary constants. Thus, the Lie algebra g of infinitesimal symmetry of Eq. (2.17) is spanned by the two vector fields:

$$V_1 = \frac{\partial}{\partial x}, \quad V_2 = (\nu - 2\kappa)t\frac{\partial}{\partial t} + \alpha(\nu-\kappa)x\frac{\partial}{\partial x} + \alpha u\frac{\partial}{\partial u}.$$

For the symmetry of V_2, corresponding characteristic equation is

$$\frac{dt}{(\nu-2\kappa)t} = \frac{dx}{\alpha(\nu-\kappa)x} = \frac{du}{\alpha u},$$

from which its solving yields the similarity variables

$$\zeta = xt^{\frac{\alpha(\kappa-\nu)}{\nu-2\kappa}}, \quad u(x,t) = t^{\frac{\alpha}{\nu-2\kappa}}\mathcal{F}(\zeta). \tag{2.19}$$

Now, from the following theorem we reduce the FPDE (2.17) to an FODE.

Theorem 12 *The transformation (2.19) reduces (2.17) to the following non-linear ordinary differential equation of fractional order:*

$$\left(\mathcal{P}^{1-\alpha+\frac{\alpha}{\nu-2\kappa},\alpha}_{\frac{\nu-2\kappa}{\alpha(\nu-\kappa)}}\mathcal{F}\right)(\zeta) = (\kappa+1)\mathcal{F}^{\kappa}\mathcal{F}' + (\nu^2+\nu)\mathcal{F}^{\nu-1}\left(\mathcal{F}'\right)^2 + (\nu+1)\mathcal{F}^{\nu}\mathcal{F}''$$

with the Erdélyi-Kober fractional differential operator $\mathcal{P}^{\tau,\alpha}_{\beta}$ of order:

$$\left(\mathcal{P}^{\tau,\alpha}_{\beta}\mathcal{F}\right) := \prod_{j=0}^{n-1}\left(\tau + j - \frac{1}{\beta}\zeta\frac{d}{d\zeta}\right)\left(\mathcal{K}^{\tau+\alpha,n-\alpha}_{\beta}\mathcal{F}\right)(\zeta), \; n = \begin{cases} [\alpha]+1, & \alpha \notin \mathbb{N} \\ \alpha, & \alpha \in \mathbb{N} \end{cases}$$

where

$$\left(\mathcal{K}^{\tau,\alpha}_{\beta}\mathcal{F}\right)(\zeta) := \begin{cases} \frac{1}{\Gamma(\alpha)}\int_1^{\infty}(u-1)^{\alpha-1}u^{-(\tau+\alpha)}\mathcal{F}(\zeta u^{\frac{1}{\beta}})du, \\ \mathcal{F}(\zeta), \quad \alpha = 0, \end{cases}$$

is the Erdélyi-Kober fractional integral operator.

Proof: Let $n-1 < \alpha < n, \quad n = 1,2,3,\ldots.$ Based on the Riemann-Liouville fractional derivative, one can have

$$\frac{\partial^{\alpha}u}{\partial t^{\alpha}} = \frac{\partial^n}{\partial t^n}\left[\frac{1}{\Gamma(n-\alpha)}\int_0^t (t-s)^{n-\alpha-1}s^{\frac{\alpha}{\nu-2\kappa}}\mathcal{F}\left(xs^{\frac{\alpha(\kappa-\nu)}{\nu-2\kappa}}\right)ds\right]. \tag{2.20}$$

Letting $v = \frac{t}{s}$, one can get $ds = -\frac{t}{v^2}dv$; therefore (2.20) can be written as

$$\frac{\partial^{\alpha}u}{\partial t^{\alpha}} = \frac{\partial^n}{\partial t^n}\left[t^{\frac{(n-\alpha)(\nu-2\kappa)+\alpha}{\nu-2\kappa}}\left(\mathcal{K}^{1+\frac{\alpha}{\nu-2\kappa},n-\alpha}_{\frac{\nu-2\kappa}{\alpha(\nu-\kappa)}}\mathcal{F}\right)(\zeta)\right].$$

Taking into account the relation $(\zeta = xt^{\frac{\alpha(\kappa-\nu)}{\nu-2\kappa}})$, we can obtain

$$t\frac{\partial}{\partial t}\phi(\zeta) = t\frac{\partial\zeta}{\partial t}\frac{d\phi(\zeta)}{d\zeta} = \frac{\alpha(\kappa-\nu)}{\nu-2\kappa}\zeta\frac{d\phi(\zeta)}{d\zeta}.$$

Therefore one can get

$$\begin{aligned}
&\frac{\partial^n}{\partial t^n}\left[t^{\frac{(n-\alpha)(\nu-2\kappa)+\alpha}{\nu-2\kappa}}\left(\mathcal{K}^{1+\frac{\alpha}{\nu-2\kappa},n-\alpha}_{\frac{\nu-2\kappa}{\alpha(\nu-\kappa)}}\mathcal{F}\right)(\zeta)\right]\\
&=\frac{\partial^{n-1}}{\partial t^{n-1}}\left[\frac{\partial}{\partial t}\left(t^{n-\alpha+\frac{\alpha}{\nu-2\kappa}}\left(\mathcal{K}^{1+\frac{\alpha}{\nu-2\kappa},n-\alpha}_{\frac{\nu-2\kappa}{\alpha(\nu-\kappa)}}\mathcal{F}\right)(\zeta)\right)\right]\\
&=\frac{\partial^{n-1}}{\partial t^{n-1}}\left[t^{n-\alpha+\frac{\alpha}{\nu-2\kappa}-1}\left(n-\alpha+\frac{\alpha}{\nu-2\kappa}+\frac{\alpha(\kappa-\nu)}{\nu-2\kappa}\zeta\frac{d}{d\zeta}\right)\left(\mathcal{K}^{1+\frac{\alpha}{\nu-2\kappa},n-\alpha}_{\frac{\nu-2\kappa}{\alpha(\nu-\kappa)}}\mathcal{F}\right)(\zeta)\right]\\
&=\ldots=t^{-\alpha+\frac{\alpha}{\nu-2\kappa}}\prod_{j=0}^{n-1}\left(1-\alpha+\frac{\alpha}{\nu-2\kappa}+j+\frac{\alpha(\kappa-\nu)}{\nu-2\kappa}\zeta\frac{d}{d\zeta}\right)\left(\mathcal{K}^{1+\frac{\alpha}{\nu-2\kappa},n-\alpha}_{\frac{\nu-2\kappa}{\alpha(\nu-\kappa)}}\mathcal{F}\right)(\zeta)\\
&=t^{-\alpha+\frac{\alpha}{\nu-2\kappa}}\left(\mathcal{P}^{1-\alpha+\frac{\alpha}{\nu-2\kappa},\alpha}_{\frac{\nu-2\kappa}{\alpha(\nu-\kappa)}}\mathcal{F}\right)(\zeta).
\end{aligned}$$

This completes the proof.

2.3.1 Exact solutions of time-fractional Fokker-Planck equation by invariant subspace method

From now on we make the assumption that the fractional derivative is Caputo one. We briefly describe the invariant subspace method applicable to the time FPDEs of the form [192, 40]:

$$ {}^C D_t^\alpha(u)=\Xi[u],\quad \alpha\in\mathbb{R}_+. \tag{2.21}$$

Definition 15 *A finite dimensional linear space* $\mathfrak{W}_n=span\{\omega_1(x),\omega_2(x),\ldots,\omega_n(x)\}$ *is said to be an invariant subspace with respect to* Ξ*, if* $\Xi[\mathfrak{W}_n]\subseteq\mathfrak{W}_n$.

Suppose that the Eq. (2.21) admits an invariant subspace $\mathfrak{W}_n$. Then from the above definition, there exist the expansion coefficient functions $\psi_1,\psi_2,\ldots,\psi_n$ such that

$$\Xi\left[\sum_{i=1}^n\lambda_i\omega_i(x)\right]=\sum_{i=1}^n\psi_i(\lambda_1,\lambda_2,\ldots,\lambda_n)\omega_i(x),\quad \lambda_i\in\mathbb{R}. \tag{2.22}$$

Hence

$$u(x,t)=\sum_{i=1}^n\lambda_i(t)\omega_i(x) \tag{2.23}$$

is the solution of Eq. (2.21), if the expansion coefficients $\lambda_i(t),\ i=1,\ldots,n$, satisfy a system of FODEs:

$$\begin{cases}{}^C D_t^\alpha(\lambda_1(t))=\psi_1(\lambda_1(t),\lambda_2(t),\ldots,\lambda_n(t)),\\ \qquad\vdots\\ {}^C D_t^\alpha(\lambda_n(t))=\psi_n(\lambda_1(t),\lambda_2(t),\ldots,\lambda_n(t)).\end{cases} \tag{2.24}$$

Now, in order to find the invariant subspace $\mathfrak{W}_n$ of a given fractional equation, one can use the following theorem [80].

Theorem 13 *Let functions ω_i, $i = 1, \ldots, n$ form the fundamental set of solutions of a linear n^{th} order ODE*

$$L[y] \equiv y^{(n)} + a_1(x)y^{(n-1)} + \cdots + a_{n-1}(x)y' + a_n(x)y = 0, \tag{2.25}$$

and let Ξ be a smooth enough function. Then the subspace $\mathfrak{W}_n = span\{\omega_1(x), \omega_2(x), \ldots, \omega_n(x)\}$ is invariant with respect to the operator Ξ of order $k \leq n-1$ if and only if

$$L(\Xi[y])|_{L[y]=0} = 0. \tag{2.26}$$

Moreover, dimension of the invariant subspace $\mathfrak{W}_n$ for the kth order nonlinear ODE operator $\Xi[y]$ satisfies $n \leq 2k+1$.

Let us consider some certain cases:

- $\kappa = 1,\ \nu = 0$:

In this case, Eq. (2.17) reduces into

$${}^C D_t^{\alpha}(u) = 2uu_x + u_{xx}. \tag{2.27}$$

After some calculations, we can find that $\mathfrak{W}_2 = span\{1, x\}$ is the invariant subspace of $\Xi[u] = 2uu_x + u_{xx}$, because

$$\Xi[\lambda_1 + \lambda_2 x] = 2\lambda_1\lambda_2 + 2\lambda_2^2 x \in \mathfrak{W}_2. \tag{2.28}$$

This results in an exact solution of the form:

$$u(x,t) = \lambda_1(t) + \lambda_2(t)x, \tag{2.29}$$

that is sufficient to solve the system of FODEs

$$\begin{cases} {}^C D_t^{\alpha}(\lambda_2(t)) = 2\lambda_2^2(t), \\ {}^C D_t^{\alpha}(\lambda_1(t)) = 2\lambda_1(t)\lambda_2(t). \end{cases} \tag{2.30}$$

Solving these equations implies

$$\lambda_1(t) = t^{-\alpha}, \quad \lambda_2(t) = \frac{\Gamma(1-\alpha)}{2\Gamma(1-2\alpha)} t^{-\alpha}. \tag{2.31}$$

Therefore, from Eq. (2.29) we obtain

$$u(x,t) = t^{-\alpha} + \frac{\Gamma(1-\alpha)}{2\Gamma(1-2\alpha)} t^{-\alpha} x. \tag{2.32}$$

• $\kappa = -1, \ \nu = 1$:
In this case, Eq. (2.17) reduces into

$$ {}^{C}D_t^{\alpha}(u) = 2u_x^2 + 2uu_{xx}. \tag{2.33}$$

After some calculations, we can find that $\mathfrak{W}_2 = span\{1, x^2\}$ is the invariant subspace of $\Xi[u] = 2uu_x + u_{xx}$, because

$$\Xi[\lambda_1 + \lambda_2 x^2] = 4\lambda_1\lambda_2 + 12\lambda_2^2 x^2 \in \mathfrak{W}_2. \tag{2.34}$$

This results in an exact solution of the form:

$$u(x,t) = \lambda_1(t) + \lambda_2(t)x^2, \tag{2.35}$$

that is sufficient to solve the system of FODEs

$$\begin{cases} {}^{C}D_t^{\alpha}(\lambda_2(t)) = 12\lambda_2^2(t), \\ {}^{C}D_t^{\alpha}(\lambda_1(t)) = 4\lambda_1(t)\lambda_2(t). \end{cases} \tag{2.36}$$

Solving these equations implies

$$\lambda_1(t) = t^{-\alpha}, \quad \lambda_2(t) = \frac{\Gamma(1-\alpha)}{12\Gamma(1-2\alpha)} t^{-\alpha}. \tag{2.37}$$

Therefore, from Eq. (2.35) we obtain

$$u(x,t) = t^{-\alpha} + \frac{\Gamma(1-\alpha)}{12\Gamma(1-2\alpha)} t^{-\alpha} x^2. \tag{2.38}$$

2.4 Lie symmetries of time-fractional Fisher equation

The Fisher equation suggested in [51] is a model for the spatial and temporal propagation of a virile gene in an infinite medium. This equation has many applications in flame propagation [104], nuclear reaction theory [34], chemical kinetics [126], autocatalytic chemical reaction [15], neurophysiology [174] and branching Brownian motion process [33]. The Fisher equation with fractional derivative in time sense [160] is as follows [77]:

$$\partial_t^{\alpha} u = u_{xx} + u(1-u^n), \quad n = 0, 1, 2, \ldots, \tag{2.39}$$

where $\partial_t^{\alpha} u := {}^{RL}D_t^{\alpha} u$ stands for Riemann-Liouville derivative of order $\alpha \in (0,1)$.

According to the Lie theory, applying the prolongation $Pr^{(\alpha,2)}V$ to Eq. (2.39), we can get the following invariance criterion:

$$\phi_{\alpha}^{0} - \phi^{xx} - \phi\,(1-(n+1)u^n) = 0. \tag{2.40}$$

Substituting the expressions for ϕ_α^0 and ϕ^{xx} into (2.40) and equating various powers of derivatives of u to zero we obtain an overdetermined system of linear equations as follows:

$$
\begin{aligned}
&\xi_x^1 = \xi_u^1 = \xi_t^2 = \xi_u^2 = \phi_{uu} = 0,\\
&2\xi_x^2 - \alpha\xi_t^1 = 0,\\
&\xi_{xx}^2 - 2\phi_{xu} = 0,\\
&(1-\alpha)\xi_{tt}^1 + 2\phi_{tu} = 0,\\
&(2-\alpha)\xi_{ttt}^1 + 3\phi_{ttu} = 0,\\
&-\phi_{xx} + u^n(n+1)\phi + (u - 1 - u^{n+1})\phi_u + \alpha(u^{n+1} - u)\xi_t^1 - u\partial_t^\alpha\phi_u + \partial_t^\alpha\phi = 0,\\
&\sum_{k=3}^{\infty}\left(\begin{array}{c}\alpha\\k\end{array}\right)\frac{\partial^{k+1}}{\partial t^k\partial u}\phi \times D_{t^{\alpha-k}}u + \sum_{k=3}^{\infty}\frac{1}{1+k}\Big[\left(k\left(\begin{array}{c}\alpha\\k\end{array}\right) - \alpha\right)D_{t^{\alpha-k}}u\\
&\times D_{t^{1+k}}\xi^1 - (k+1)D_{t^{\alpha-k}}u_x \times D_{t^k}\xi^2\Big] = 0.
\end{aligned}
$$

Solving these determining equations, obtained from (2.40), yields:

$$\xi^1 = 4tc_2, \quad \xi^2 = c_1 + 2\alpha x c_2, \quad \phi = c_3 u + (3\alpha - 2)uc_2 + \mathcal{C}(x,t),$$

where c_1, c_2 and c_3 are arbitrary constants and $\mathcal{C}(x,t)$ is an arbitrary solution of (2.39). Thus, the Lie algebra g of infinitesimal symmetries of Eq. (2.39) is spanned by the vector fields:

$$V_1 = \frac{\partial}{\partial x}, \quad V_2 = 4t\frac{\partial}{\partial t} + 2\alpha x\frac{\partial}{\partial x} + (3\alpha - 2)u\frac{\partial}{\partial u}, \quad V_3 = u\frac{\partial}{\partial u}, \quad V_\infty = \mathcal{C}(x,t)\frac{\partial}{\partial u}.$$

Now, we are ready to use the obtained vector fields to reduce Eq. (2.39) into FODEs. To do this, we consider some special cases as follows:

Case 1: $V = V_2$**.**

In this case, the characteristic equation corresponding to V_2 is:

$$\frac{dt}{4t} = \frac{dx}{2\alpha x} = \frac{du}{(3\alpha - 2)u},$$

which yields the similarity variables

$$u(x,t) = t^{\frac{3\alpha-2}{4}}\mathcal{F}(\zeta), \quad \zeta = xt^{\frac{-\alpha}{2}}. \tag{2.41}$$

Now, from the following theorem we reduce the time-fractional Fisher (TFF) equation (2.39) to an FODE.

Theorem 14 *Using (2.41), Eq. (2.39) is reduced to the following nonlinear FODE:*

$$\left(\mathcal{P}_{\frac{2}{\alpha}}^{-\frac{\alpha}{4}+\frac{1}{2},\alpha}\mathcal{F}\right)(\zeta) = \mathcal{F}'' + \mathcal{F}\left(1 - \mathcal{F}^n\right).$$

Proof: Based on the Riemann-Liouville fractional derivative and similarity variables related to V_2, one can have

$$\frac{\partial^\alpha u}{\partial t^\alpha} = \frac{\partial^n}{\partial t^n}\Big[\frac{1}{\Gamma(n-\alpha)}\int_0^t (t-s)^{n-\alpha-1} s^{\frac{3\alpha-2}{4}} \mathcal{F}\left(xs^{\frac{-\alpha}{2}}\right) ds\Big]. \tag{2.42}$$

Letting $\rho = \frac{t}{s}$, one can get $ds = -\frac{t}{\rho^2} d\rho$, therefore

$$\frac{\partial^\alpha u}{\partial t^\alpha} = \frac{\partial^n}{\partial t^n}\Big[t^{n-\frac{\alpha}{4}-\frac{1}{2}}\left(\mathcal{K}_{\frac{2}{\alpha}}^{\frac{3\alpha+2}{4}, n-\alpha}\mathcal{F}\right)(\zeta)\Big]. \tag{2.43}$$

Taking into account the relation ($\zeta = xt^{\frac{-\alpha}{2}}$), we can obtain

$$t\frac{\partial}{\partial t}\phi(\zeta) = t\frac{\partial \zeta}{\partial t}\frac{d\phi(\zeta)}{d\zeta} = -\frac{\alpha}{2}\zeta\frac{d\phi(\zeta)}{d\zeta}. \tag{2.44}$$

Therefore by setting $\Phi(\zeta) = \left(\mathcal{K}_{\frac{2}{\alpha}}^{\frac{3\alpha+2}{4}, n-\alpha}\mathcal{F}\right)(\zeta)$ one can get

$$\begin{aligned}
&\frac{\partial^n}{\partial t^n}\big[t^{n-\frac{\alpha}{4}-\frac{1}{2}}\Phi(\zeta)\big] = \frac{\partial^{n-1}}{\partial t^{n-1}}\Big[\frac{\partial}{\partial t}\left(t^{n-\frac{\alpha}{4}-\frac{1}{2}}\Phi(\zeta)\right)\Big] \\
&= \frac{\partial^{n-1}}{\partial t^{n-1}}\Big[t^{n-\frac{\alpha}{4}-\frac{3}{2}}\left(n-\frac{\alpha}{4}-\frac{1}{2}-\frac{\alpha}{2}\zeta\frac{d}{d\zeta}\right)\Phi(\zeta)\Big] = \ldots \\
&= t^{-\frac{\alpha}{4}-\frac{1}{2}}\prod_{j=0}^{n-1}\left(-\frac{\alpha}{4}+\frac{1}{2}+j-\frac{\alpha}{2}\zeta\frac{d}{d\zeta}\right)\Phi(\zeta) = t^{-\frac{\alpha}{4}-\frac{1}{2}}\left(\mathcal{P}_{\frac{2}{\alpha}}^{-\frac{\alpha}{4}+\frac{1}{2}, \alpha}\mathcal{F}\right)(\zeta).
\end{aligned} \tag{2.45}$$

This completes the proof.

Case 2: $V = V_1 + V_2 + V_3$.

For the symmetry of $V_1 + V_2 + V_3$, corresponding characteristic equation is

$$\frac{dt}{4t} = \frac{dx}{2\alpha x + 1} = \frac{du}{(3\alpha - 1)u},$$

which yields the similarity variables

$$u(x,t) = t^{\frac{3\alpha-1}{4}}\mathcal{F}(\zeta), \quad \zeta = \left(\frac{2\alpha x + 1}{2\alpha}\right)t^{\frac{-\alpha}{2}}. \tag{2.46}$$

As shown in previous theorem, another reduction of Eq. (2.39) to an FODE can be written as follows:

Theorem 15 *Eq. (2.39) is reducible into a nonlinear fractional ordinary differential equation of the form:*

$$\left(\mathcal{P}_{\frac{2}{\alpha}}^{\frac{3-\alpha}{4}, \alpha}\mathcal{F}\right)(\zeta) = \mathcal{F}'' + \mathcal{F}\left(1 - \mathcal{F}^n\right), \tag{2.47}$$

using the transformation (2.46).

Proof: Similar to the previous case, we have

$$\frac{\partial^\alpha u}{\partial t^\alpha} = \frac{\partial^n}{\partial t^n}\Big[\frac{1}{\Gamma(n-\alpha)}\int_0^t (t-s)^{n-\alpha-1} s^{\frac{3\alpha-1}{4}} \mathcal{F}\left(\frac{2\alpha x+1}{2\alpha} s^{\frac{-\alpha}{2}}\right) ds\Big].$$

With the same change of variable in the previous theorem, we find the following equality:

$$\frac{\partial^\alpha u}{\partial t^\alpha} = \frac{\partial^n}{\partial t^n}\Big[t^{n-\frac{\alpha-1}{4}}\left(\mathcal{K}_{\frac{2}{\alpha}}^{\frac{3\alpha+1}{2},n-\alpha}\mathcal{F}\right)(\zeta)\Big].$$

From (2.46) we obtain

$$t\frac{\partial}{\partial t}\phi(\zeta) = t\frac{\partial\zeta}{\partial t}\frac{d\phi(\zeta)}{d\zeta} = -\frac{\alpha}{2}\zeta\frac{d\phi(\zeta)}{d\zeta}.$$

Therefore by setting $\Phi(\zeta) = \left(\mathcal{K}_{\frac{2}{\alpha}}^{\frac{3\alpha+1}{2},n-\alpha}\mathcal{F}\right)(\zeta)$ and keeping on the similar process of the previous theorem we find the desired result.

Case 3: $V = V_1 + \rho V_3$.

Characteristic equation in this case can be written as

$$\frac{dt}{0} = \frac{dx}{1} = \frac{du}{\rho u},$$

from which solving them concludes the similarity variables of the form

$$u(x,t) = e^{\rho x}\mathcal{F}(t),$$

where $\mathcal{F}(t)$ satisfies the following equation:

$$^{RL}D_t^\alpha \mathcal{F}(t) = \rho^2\mathcal{F}(t) + \mathcal{F}(t)\left(1 - e^{n\rho x}\mathcal{F}^n(t)\right). \tag{2.48}$$

The trivial solution of Eq. (2.48) can be derived for $\rho \neq 0$, but for $\rho = 0$ Eq. (2.48) reduces to

$$^{RL}D_t^\alpha \mathcal{F}(t) = \mathcal{F}(t)\left(1 - \mathcal{F}^n(t)\right).$$

Finding the exact solution of the above equation is cumbersome but for $n = 0$ for which we have

$$^{RL}D_t^\alpha \mathcal{F}(t) = 0,$$

the invariant solution is

$$u(x,t) = C_1 t^{\alpha-1}.$$

Also, for $n = 0$ and $\rho \neq 0$ that Eq. (2.39) reduces to the linear anomalous diffusion equation, we have transformed the form of Eq. (2.48) as follows:

$$^{RL}D_t^\alpha \mathcal{F}(t) = \rho^2\mathcal{F}(t). \tag{2.49}$$

The solution of Eq. (2.49) can be written as

$$\mathcal{F}(t) = t^{\alpha-1}E_{\alpha,\alpha}(\rho^2 t^\alpha).$$

Therefore an invariant solution of Eq. (2.39) with $n = 0$ can be written as

$$u(x,t) = e^{\rho x}t^{\alpha-1}E_{\alpha,\alpha}(\rho^2 t^\alpha).$$

2.5 Lie symmetries of time-fractional $K(m,n)$ equation

This section mainly focuses on the Lie group analysis of time-fractional $K(m,n)$ equation [176, 148]:

$$\partial_t^\alpha u - a\left(u^n\right)_x + b\left(u^m\right)_{xxx} = 0. \tag{2.50}$$

The well-known $K(m,n)$ equation, which is a generalization of the KdV equation, describes the evolution of weakly nonlinear and weakly dispersive waves used in various fields such as solid-state physics, plasma physics, fluid physics and quantum field theory. After Odibat [148], which introduced the Eq. (2.50) and considered three special cases $K(2,2)$, $K(3,3)$ and $K(n,n)$, Koçak *et al.* in [113] considered the nonlinear dispersive $K(m,n,1)$ equations with fractional time derivatives by using the homotopy perturbation method.

According to the Lie theory, applying the prolongation $Pr^{(\alpha,3)}V$ to Eq. (2.50), we can get the following invariance criterion:

$$\begin{aligned}
&\phi_\alpha^0 - an(n-1)\phi u^{n-2}u_x - anu^{n-1}\phi^x + bm(m-1)(m-2)(m-3)u^{m-4}\phi u_x^3\\
&+ 3bm(m-1)(m-2)u^{m-3}u_x^2\phi^x + 3bm(m-1)(m-2)u^{m-3}\phi u_x u_{xx}\\
&+ 3bm(m-1)u^{m-2}\phi^x u_{xx} + 3bm(m-1)u^{m-2}u_x\phi^{xx}\\
&+ bm(m-1)u^{m-2}\phi u_{xxx} + bmu^{m-1}\phi^{xxx} = 0.
\end{aligned} \tag{2.51}$$

Then, we obtain the following forms of the coefficient functions:

$$\xi^1 = (m+2-3n)tc_2, \quad \xi^2 = c_1 + \alpha(m-n)xc_2, \quad \phi = 2\alpha uc_2,$$

where c_1 and c_2 are arbitrary constants. Thus, the Lie algebra g of infinitesimal symmetry of Eq. (2.50) is spanned by the two vector fields:

$$V_1 = \frac{\partial}{\partial x}, \quad V_2 = (m+2-3n)t\frac{\partial}{\partial t} + \alpha(m-n)x\frac{\partial}{\partial x} + 2\alpha u\frac{\partial}{\partial u}.$$

For the symmetry of V_2, corresponding characteristic equation is

$$\frac{dt}{(m+2-3n)t} = \frac{dx}{\alpha(m-n)x} = \frac{du}{2\alpha u},$$

from which its solving yields the similarity variables

$$\zeta = xt^{\frac{\alpha(n-m)}{m+2-3n}}, \quad u(x,t) = t^{\frac{2\alpha}{m+2-3n}}\mathcal{F}(\zeta). \tag{2.52}$$

Now, from the following theorem we reduce the FPDE (2.50) to an FODE.

Theorem 16 *The transformation (2.52) reduces (2.50) to the following nonlinear ODE of fractional order:*

$$\left(\mathcal{P}^{1-\alpha+\frac{2\alpha}{m+2-3n},\alpha}_{\frac{m+2-3n}{\alpha(m-n)}}\mathcal{F}\right)(\zeta) = an\mathcal{F}^{n-1}\mathcal{F}' - bm\mathcal{F}^{m-1}\mathcal{F}'''$$
$$- bm(m-1)(m-2)\mathcal{F}^{m-3}\left(\mathcal{F}'\right)^3 - 3bm(m-1)\mathcal{F}^{m-2}\mathcal{F}'\mathcal{F}''.$$

Proof: Let $k-1<\alpha<k,\ \ k=1,2,3,....$ Based on the Riemann-Liouville fractional derivative, one can have

$$\frac{\partial^\alpha u}{\partial t^\alpha} = \frac{\partial^k}{\partial t^k}\left[\frac{1}{\Gamma(k-\alpha)} \times \int_0^t (t-s)^{k-\alpha-1} s^{\frac{2\alpha}{m+2-3n}} \mathcal{F}\left(xs^{\frac{\alpha(n-m)}{m+2-3n}}\right) ds\right]. \quad (2.53)$$

Letting $v=\frac{t}{s}$, one can get $ds=-\frac{t}{v^2}dv$; therefore (2.53) can be written as

$$\frac{\partial^\alpha u}{\partial t^\alpha} = \frac{\partial^k}{\partial t^k}\left[t^{k-\alpha+\frac{2\alpha}{m+2-3n}}\left(\mathcal{K}^{1+\frac{2\alpha}{m+2-3n},k-\alpha}_{\frac{m+2-3n}{\alpha(m-n)}}\mathcal{F}\right)(\zeta)\right].$$

Taking into account the relation ($\zeta = xt^{\frac{\alpha(n-m)}{m+2-3n}}$), we can obtain

$$t\frac{\partial}{\partial t}\phi(\zeta) = t\frac{\partial\zeta}{\partial t}\frac{d\phi(\zeta)}{d\zeta} = \frac{\alpha(n-m)}{m+2-3n}\zeta\frac{d\phi(\zeta)}{d\zeta}.$$

Therefore one can get

$$\begin{aligned}
&\frac{\partial^k}{\partial t^k}\left[t^{k-\alpha+\frac{\alpha}{m+2-3n}}\left(\mathcal{K}^{1+\frac{2\alpha}{m+2-3n},k-\alpha}_{\frac{m+2-3n}{\alpha(m-n)}}\mathcal{F}\right)(\zeta)\right]\\
&= \frac{\partial^{k-1}}{\partial t^{k-1}}\left[\frac{\partial}{\partial t}\left(t^{k-\alpha+\frac{2\alpha}{m+2-3n}}\left(\mathcal{K}^{1+\frac{2\alpha}{m+2-3n},k-\alpha}_{\frac{m+2-3n}{\alpha(m-n)}}\mathcal{F}\right)(\zeta)\right)\right]\\
&= \frac{\partial^{k-1}}{\partial t^{k-1}}\left[t^{k-\alpha+\frac{2\alpha}{m+2-3n}-1}\left(k-\alpha+\frac{2\alpha}{m+2-3n}\right.\right.\\
&\left.\left.+\frac{\alpha(n-m)}{m+2-3n}\zeta\frac{d}{d\zeta}\right)\left(\mathcal{K}^{1+\frac{2\alpha}{m+2-3n},k-\alpha}_{\frac{m+2-3n}{\alpha(m-n)}}\mathcal{F}\right)(\zeta)\right]\\
&= \ldots\\
&= t^{-\alpha+\frac{2\alpha}{m+2-3n}}\prod_{j=0}^{k-1}\left(1-\alpha+\frac{2\alpha}{m+2-3n}+j\right.\\
&\left.+\frac{\alpha(n-m)}{m+2-3n}\zeta\frac{d}{d\zeta}\right)\left(\mathcal{K}^{1+\frac{2\alpha}{m+2-3n},k-\alpha}_{\frac{m+2-3n}{\alpha(m-n)}}\mathcal{F}\right)(\zeta)\\
&= t^{-\alpha+\frac{2\alpha}{m+2-3n}}\left(\mathcal{P}^{1-\alpha+\frac{2\alpha}{m+2-3n},\alpha}_{\frac{m+2-3n}{\alpha(m-n)}}\mathcal{F}\right)(\zeta).
\end{aligned}$$

This ends the proof.

2.6 Lie symmetries of time-fractional gas dynamics equation

The time-fractional gas dynamics (TFGD) equation [75] has the form

$$\partial_t^\alpha u + \frac{1}{2}\left(u^2\right)_x - u\left(1-u\right) = 0,\ 0 < \alpha \leq 1, \tag{2.54}$$

where $u(x,t)$ is the probability density and $\partial_t^\alpha u := {}^{RL}D_t^\alpha u$ stands for Riemann-Liouville derivative of order α. When $\alpha = 1$, the TFGD equation reduces to the classical gas dynamics equation which is considered as a case study for solving hyperbolic conservation laws because it depicts the next level of complexity after the Burger's equation.

According to the Lie theory, applying the prolongation $Pr^{(\alpha,1)}V$ to Eq. (2.54), we can get the following invariance criterion:

$$\phi_\alpha^0 - \phi u_x + u\phi^x - \phi + 2\phi u = 0. \tag{2.55}$$

Substituting (2.8) into (2.55), and equating the coefficients of the various monomials in partial derivatives with respect to x and various powers of u, one can find the determining equations for the symmetry group of Eq. (2.54). Solving these equations, we obtain the following forms of the coefficient functions:

$$\xi^1 = tc_2, \quad \xi^2 = c_1 + 2\alpha x c_2, \quad \phi = \alpha u c_2, \tag{2.56}$$

where c_1 and c_2 are arbitrary constants. Thus, the Lie algebra g of infinitesimal symmetry of Eq. (2.54) is spanned by the two vector fields:

$$V_1 = \frac{\partial}{\partial x}, \quad V_2 = t\frac{\partial}{\partial t} + 2\alpha x\frac{\partial}{\partial x} + \alpha u\frac{\partial}{\partial u}. \tag{2.57}$$

For the symmetry of V_2, corresponding characteristic equation is

$$\frac{dt}{t} = \frac{dx}{2\alpha x} = \frac{du}{\alpha u}, \tag{2.58}$$

from which its solving yields the similarity variables

$$\zeta = xt^{-2\alpha}, \quad u(x,t) = t^\alpha \mathcal{F}(\zeta). \tag{2.59}$$

Now, from the following theorem we reduce the TFGD (2.54) to an FODE.

Theorem 17 *The transformation (2.59) reduces (2.54) to the following nonlinear ordinary differential equation of fractional order:*

$$\left(\mathcal{P}_{\frac{1}{2\alpha}}^{1,\alpha}\mathcal{F}\right)(\zeta) - \mathcal{F}(\mathcal{F}' + \mathcal{F} - 1) = 0. \tag{2.60}$$

Proof: Let $n-1<\alpha<n, \quad n=1,2,3,....$ Based on the Riemann-Liouville fractional derivative, one can have

$$\frac{\partial^\alpha u}{\partial t^\alpha}=\frac{\partial^n}{\partial t^n}\Big[\frac{1}{\Gamma(n-\alpha)}\int_0^t (t-s)^{n-\alpha-1}s^\alpha \mathcal{F}\left(xs^{-2\alpha}\right)ds\Big]. \tag{2.61}$$

Letting $v=\frac{t}{s}$, one can get $ds=-\frac{t}{v^2}dv$; therefore (2.61) can be written as

$$\frac{\partial^\alpha u}{\partial t^\alpha}=\frac{\partial^n}{\partial t^n}\Big[t^n\left(\mathcal{K}^{1+\alpha,n-\alpha}_{\frac{1}{2\alpha}}\mathcal{F}\right)(\zeta)\Big].$$

Taking into account the relation ($\zeta=xt^{-2\alpha}$), we can obtain

$$t\frac{\partial}{\partial t}\phi(\zeta)=t\frac{\partial\zeta}{\partial t}\frac{d\phi(\zeta)}{d\zeta}=-2\alpha\zeta\frac{d\phi(\zeta)}{d\zeta}.$$

Therefore one can get

$$\begin{aligned}
&\frac{\partial^n}{\partial t^n}\Big[t^n\left(\mathcal{K}^{1+\alpha,n-\alpha}_{\frac{1}{2\alpha}}\mathcal{F}\right)(\zeta)\Big]=\frac{\partial^{n-1}}{\partial t^{n-1}}\Big[\frac{\partial}{\partial t}\left(t^n\left(\mathcal{K}^{1+\alpha,n-\alpha}_{\frac{1}{2\alpha}}\mathcal{F}\right)(\zeta)\right)\Big]\\
&=\frac{\partial^{n-1}}{\partial t^{n-1}}\Big[t^{n-1}\left(n-2\alpha\zeta\frac{d}{d\zeta}\right)\left(\mathcal{K}^{1+\alpha,n-\alpha}_{\frac{1}{2\alpha}}\mathcal{F}\right)(\zeta)\Big]\\
&=...=\prod_{j=0}^{n-1}\left(1+j-2\alpha\zeta\frac{d}{d\zeta}\right)\left(\mathcal{K}^{1+\alpha,n-\alpha}_{\frac{1}{2\alpha}}\mathcal{F}\right)(\zeta)=\left(\mathcal{P}^{1,\alpha}_{\frac{1}{2\alpha}}\mathcal{F}\right)(\zeta).
\end{aligned}$$

This finishes the proof.

2.7 Lie symmetries of time-fractional diffusion-absorption equation

A pioneering investigation of localization-extinction phenomena nonlinear degenerate parabolic PDEs was first performed by Kersner in the 1960s-70s. Key results about these PDEs, including equations from diffusion-absorption theory, are reflected by Kalashnikov in [103]. Among the discussed models of Kalashnikov, the diffusion-absorption equation with the critical absorption exponent:

$$v_t=\left(v^\sigma v_x\right)_x-v^{1-\sigma} \tag{2.62}$$

is a famous one, where σ is a positive parameter. It is well known in filtration theory that the terms $-v^{1-\sigma}$ show the absorption and describe the seepage on a permeable bed. Some explicit localized solutions of Eq. (2.62) were reported in [167]. From the filtration theory, by introducing $u=v^\sigma$, as a pressure variable, we obtain a PDE with quadratic differential operator and constant sink

$$u_t = uu_{xx} + \frac{1}{\sigma}(u_x)^2 - \sigma. \tag{2.63}$$

Time-fractional version of Eq. (2.63) has the following form [81]:

$$\partial_t^\alpha u = uu_{xx} + \frac{1}{\sigma}(u_x)^2 - \sigma, \quad \alpha \in (0,1). \tag{2.64}$$

An overdetermined system of partial linear differential equations can be extracted from applying $Pr^{(\alpha,2)}V$ to Eq. (2.64) as follows:

$$\begin{aligned}
&\xi_u^1 = \xi_x^1 = \xi_t^2 = \xi_u^2 = \phi_{uu} = 0,\\
&-\phi + 2u\xi_x^2 - \alpha u\xi_t^1 = 0,\\
&2\sigma u\phi_{xu} - \sigma u\xi_{xx}^2 + 2\phi_x = 0,\\
&(1-\alpha)\xi_{tt}^1 + 2\phi_{tu} = 0,\\
&(2-\alpha)\xi_{ttt}^1 + 3\phi_{ttu} = 0,\\
&\phi_u + \alpha\xi_t^1 - 2\xi_x^2 = 0,\\
&-u\phi_{xx} - \sigma\phi_u + \alpha\sigma\xi_t^1 - u\partial_t^\alpha\phi_u + \partial_t^\alpha\phi = 0,\\
&\sum_{k=3}^{\infty}\left(\begin{array}{c}\alpha\\ k\end{array}\right)\frac{\partial^{k+1}}{\partial t^k\partial u}\phi \times D_{t^{\alpha-k}}u - \sum_{k=3}^{\infty}\frac{\left(\begin{array}{c}\alpha\\ k\end{array}\right)}{1+k}\Big[(k-\alpha)\,D_{t^{\alpha-k}}u\\
&\times D_{t^{1+k}}\xi^1 + (k+1)D_{t^{\alpha-k}}u_x \times D_{t^k}\xi^2\Big] = 0.
\end{aligned} \tag{2.65}$$

Solving Eqs. (2.65), we obtain the following infinitesimals:

$$\xi^1 = c_1 + tc_2, \quad \xi^2 = c_3 + xc_4, \quad \phi = 2uc_4 - \alpha uc_2,$$

where c_1, c_2, c_3 and c_4 are arbitrary constants and therefore:

$$V_1 = \frac{\partial}{\partial x}, \quad V_2 = t\frac{\partial}{\partial t} - \alpha u\frac{\partial}{\partial u}, \quad V_3 = x\frac{\partial}{\partial x} + 2u\frac{\partial}{\partial u}. \tag{2.66}$$

Let us to consider the invariant solution of Eq. (2.64) corresponding to the third vector field V_3. The similarity variable and similarity transformation corresponding to V_3 take the following form:

$$u(x,t) = x^2\mathfrak{F}(\zeta), \ \zeta = t, \tag{2.67}$$

where the function $\mathfrak{F}(\zeta)$ satisfies the following FODE:

$$x^2\left(\partial_t^\alpha\mathfrak{F}(\zeta) - 2\mathfrak{F}^2(\zeta) - \frac{4}{\sigma}\mathfrak{F}^2(\zeta)\right) = -\sigma. \tag{2.68}$$

Obviously, finding a general solution for Eq. (2.68) is not possible and we have to solve it in a restricted domain. To do this, we impose the following condition to Eq. (2.67):

$$\partial_t^\alpha \mathfrak{F}(\zeta) - \left(2 + \frac{4}{\sigma}\right) \mathfrak{F}^2(\zeta) = -\zeta^{-2\alpha}. \tag{2.69}$$

Corresponding exact solution of Eq. (2.69) has the following form:

$$\mathfrak{F}(\zeta) = \frac{\sigma\Gamma(1-\alpha) + \sqrt{\sigma^2\Gamma(1-\alpha)^2 + (8\sigma^2 + 16\sigma)\Gamma(1-2\alpha)^2}}{(\sigma+2)\Gamma(1-2\alpha)} \zeta^{-\alpha}.$$

Therefore, the invariant solution of Eq. (2.64) is

$$u(x,t) = \frac{\sigma\Gamma(1-\alpha) + \sqrt{\sigma^2\Gamma(1-\alpha)^2 + (8\sigma^2 + 16\sigma)\Gamma(1-2\alpha)^2}}{(\sigma+2)\Gamma(1-2\alpha)} t^{-\alpha} x^2,$$

provided that

$$x^2 t^{-2\alpha} = \sigma.$$

2.7.1 Exact solutions of time-fractional diffusion-absorption by invariant subspace method

Now, from Eqs. (2.64), (2.21) and Theorem 13 one can find that the dimension of invariant subspace $\mathfrak{W}_n$ for the operator $\Xi[u]$, corresponding to Eq. (2.64), satisfies $n \leq 2(2) + 1 = 5$. After some calculations, we can find that $\mathfrak{W}_2 = span\{1, x^2\}$ is the invariant subspace of $\Xi[u] = uu_{xx} + \frac{1}{\sigma}(u_x)^2 - \sigma$, because

$$\Xi[\lambda_1 + \lambda_2 x^2] = \psi_1(\lambda_1, \lambda_2) + \psi_2(\lambda_1, \lambda_2)x^2 = 2\lambda_1\lambda_2 - \sigma + 2\left(1 + \frac{2}{\sigma}\right)\lambda_2^2 x^2 \in \mathfrak{W}_2.$$

Therefore, in order to find the exact solution of the form:

$$u(x,t) = \lambda_1(t) + \lambda_2(t)x^2, \tag{2.70}$$

it is sufficient to solve the system of FODEs

$$\begin{cases} \partial_t^\alpha \lambda_1(t) = \psi_1(\lambda_1(t), \lambda_2(t)) = 2\lambda_1(t)\lambda_2(t) - \sigma, \\ \partial_t^\alpha \lambda_2(t) = \psi_2(\lambda_1(t), \lambda_2(t)) = 2\left(1 + \frac{2}{\sigma}\right)\lambda_2^2(t). \end{cases} \tag{2.71}$$

Solving the second equation of (2.71) concludes

$$\lambda_2(t) = \frac{\Gamma(1-\alpha)t^{-\alpha}}{2\left(1 + \frac{2}{\sigma}\right)\Gamma(1-2\alpha)}.$$

Substituting $\lambda_2(t)$ in the first equation of (2.71) and solving the obtained equation yields

$$\lambda_1(t) = \frac{\sigma(\sigma+2)\Gamma(1-2\alpha)t^\alpha}{\sigma\Gamma(1-\alpha) - (\sigma+2)\Gamma(\alpha+1)\Gamma(1-2\alpha)}.$$

Therefore, from Eq. (2.70) we obtain

$$u(x,t) = \frac{\sigma(\sigma+2)\Gamma(1-2\alpha)t^{\alpha}}{\sigma\Gamma(1-\alpha)-(\sigma+2)\Gamma(\alpha+1)\Gamma(1-2\alpha)} + \frac{\Gamma(1-\alpha)t^{-\alpha}}{2\left(1+\frac{2}{\sigma}\right)\Gamma(1-2\alpha)}x^2.$$

2.8 Lie symmetries of time-fractional Clannish Random Walker's parabolic equation

The time-fractional Clannish Random Walker's (CRW) parabolic equation [43, 180]:

$$\partial_t^{\alpha} u - u_x + 2uu_x + u_{xx} = 0, \tag{2.72}$$

where $\partial_t^{\alpha} u := {}^{RL}D_t^{\alpha} u$ stands for Riemann-Liouville derivative with $\alpha \in (0,1)$, is formulated for the motion of the two interacting populations which tend to be clannish, which means they want to live near those of their own kind. Some numerical and analytical methods have been utilized for this equation by many authors in literature, e.g. [175].

Now, we utilize the developed Lie symmetry method to the time-fractional CRW equation. According to the Lie theory, determining equations of the time-fractional CRW equation can be extracted from applying $Pr^{(\alpha,2)}V$ to Eq. (2.72) as follows:

$$\begin{aligned}
&\xi_u^1 = \xi_x^1 = \xi_t^2 = \xi_u = \phi_{uu} = 0,\\
&(1-\alpha)\xi_{tt}^1 + 2\phi_{tu} = 0,\\
&(2-\alpha)\xi_{ttt}^1 + 3\phi_{ttu} = 0,\\
&\alpha\xi_t^1 - 2\xi_x^2 = 0,\\
&(1-2u)\xi_x^2 - \xi_{xx}^2 - (1-2u)\alpha\xi_t^1 + 2\phi + 2\phi_{xu} = 0,\\
&\phi_{xx} - (1-2u)\phi_x - u\partial_t^{\alpha}\phi_u + \partial_t^{\alpha}\phi = 0,\\
&\sum_{k=3}^{\infty}\binom{\alpha}{k}\frac{\partial^{k+1}}{\partial t^k \partial u}\phi \times D_{t^{\alpha-k}}u + \sum_{k=3}^{\infty}\frac{\binom{\alpha}{k}}{1+k}\Big[(k-\alpha)D_{t^{\alpha-k}}u \times D_{t^{1+k}}\xi^1\\
&+(k+1)D_{t^{\alpha-k}}u_x \times D_{t^k}\xi^2\Big] = 0.
\end{aligned} \tag{2.73}$$

Solving Eqs. (2.73) and applying the condition (2.8), we obtain the following infinitesimals:

$$\xi^1 = c_1 + 4tc_2, \quad \xi^2 = c_3 + 2\alpha x c_2, \quad \phi = c_2\alpha(1-2u), \tag{2.74}$$

where c_1, c_2, c_3 and c_4 are arbitrary constants. Hence, the Lie algebra g of infinitesimal symmetries of Eq. (2.72) by utilizing

$$\xi^1(x, t, u)|_{t=0} = 0$$

is spanned by the vector fields:

$$V_1 = \frac{\partial}{\partial x}, \quad V_2 = 4t\frac{\partial}{\partial t} + 2\alpha x\frac{\partial}{\partial x} + \alpha(1 - 2u)\frac{\partial}{\partial u}.$$

These obtained generators will help us to construct the conserved vectors of the time-fractional CRW equation.

2.8.1 Exact solutions of time-fractional Clannish Random Walker's equation by invariant subspace method

Let us consider the exact solution of Eq. (2.72) by the invariant subspace method.

From Eq. (2.72) and Theorem 13 one can find that the dimension of invariant subspace $\mathfrak{W}_n$ for the operator $\Xi[u]$, corresponding to Eq. (2.72), satisfies $n \leq 2(2) + 1 = 5$. After some calculations, we can find that $\mathfrak{W}_2 = span\{1, x\}$ is the invariant subspace of $\Xi[u] = u_{xx} + 2uu_x - u_x$, because

$$\Xi[\lambda_1 + \lambda_2 x] = \psi_1(\lambda_1, \lambda_2) + \psi_2(\lambda_1, \lambda_2)x = 2\lambda_1\lambda_2 - \lambda_2 + 2\lambda_2^2 x \in \mathfrak{W}_2.$$

Therefore, in order to find the exact solution of

$$u(x, t) = \lambda_1(t) + \lambda_2(t)x, \tag{2.75}$$

it is sufficient to solve the system of FODEs

$$\begin{cases} \partial_t^\alpha \lambda_1(t) = \psi_1(\lambda_1(t), \lambda_2(t)) = 2\lambda_1(t)\lambda_2(t) - \lambda_2(t), \\ \partial_t^\alpha \lambda_2(t) = \psi_2(\lambda_1(t), \lambda_2(t)) = 2\lambda_2^2(t). \end{cases} \tag{2.76}$$

Solving the second equation of (2.76) yields

$$\lambda_2(t) = \frac{(2\alpha - 1)\Gamma(1 - \alpha)t^{-\alpha}}{2\Gamma(2 - 2\alpha)}.$$

Substituting $\lambda_2(t)$ in the first equation of (2.76) and solving the obtained equation yields

$$\lambda_1(t) = \frac{(2\alpha - 1)\Gamma(1 - \alpha)^2}{2\Gamma(2 - 2\alpha) - 2(2\alpha - 1)\Gamma(1 - \alpha)^2}.$$

Therefore, from Eq. (2.75) we obtain

$$u(x, t) = \frac{(2\alpha - 1)\Gamma(1 - \alpha)^2}{2\Gamma(2 - 2\alpha) - 2(2\alpha - 1)\Gamma(1 - \alpha)^2} + \frac{(2\alpha - 1)\Gamma(1 - \alpha)t^{-\alpha}}{2\Gamma(2 - 2\alpha)}x.$$

Also, in Eq. (2.76), if we consider the integer order differential equations, then we can find corresponding exact solution.

2.9 Lie symmetries of the time-fractional Kompaneets equation

In this section, we consider four time-fractional generalizations of the Kompaneets equation [64]. In a dimensionless form, the classical Kompaneets equation derived by Kompaneets [114] and Weymann [181] can be written as

$$f_t = \frac{1}{x^2} D_x \left[x^4 \left(f_x + f + f^2 \right) \right]. \tag{2.77}$$

This equation describes the evolution of the density function f of the energy of photons due to Compton scattering in the homogeneous fully ionized plasma.

The time-fractional generalizations of the Kompaneets equation have the following form

$$f_t = \frac{1}{x^2} J_t^\alpha D_x \left[x^4 \left(f_x + f + f^2 \right) \right], \tag{2.78}$$

$$f_t = \frac{1}{x^2} D_t^{1-\alpha} D_x \left[x^4 \left(f_x + f + f^2 \right) \right], \tag{2.79}$$

$$f_t = \frac{1}{x^2} D_x \left[x^4 \left(J_t^\alpha f_x + J_t^\alpha f + (J_t^\alpha f)^2 \right) \right], \tag{2.80}$$

$$f_t = \frac{1}{x^2} D_x \left[x^4 \left(D_t^{1-\alpha} f_x + D_t^{1-\alpha} f + (D_t^{1-\alpha} f)^2 \right) \right]. \tag{2.81}$$

All these equations can be rewritten so that their right-hand sides will be exactly the same as the right-hand side of the classical Kompaneets Eq. (2.77). Indeed, if we act on the Eq. (2.78) by the operator D_t^α of fractional differentiation and denote the dependent variable f by u, then we get the equation

$$D_t^\alpha u_t = \frac{1}{x^2} D_x \left[x^4 \left(u_x + u + u^2 \right) \right]. \tag{2.82}$$

Integrating the Eq. (2.79) with respect to t we obtain

$$f(t,x) - f(0,x) = \frac{1}{x^2} J_t^\alpha D_x \left[x^4 \left(f_x + f + f^2 \right) \right]. \tag{2.83}$$

As above, we act on both sides of this equation by the same operator D_t^α of fractional differentiation and denote the dependent variable f by u. Then we get the equation

$$^{c}D_t^\alpha u = \frac{1}{x^2} D_x \left[x^4 \left(u_x + u + u^2 \right) \right], \tag{2.84}$$

where

$$^{c}D_t^{\alpha}u = J_t^{1-\alpha}u_t = \frac{1}{\Gamma(1-\alpha)}\int_0^t \frac{u_\tau(\tau,x)}{(t-\tau)^{\alpha}}d\tau$$

is the left-sided time-fractional derivative of the Caputo type of order $\alpha \in (0,1)$ [110, 108].

In the Eq. (2.80) we can introduce a new nonlocal dependent variable $u = J_t^{\alpha}f$. Then this equation takes the form

$$D_t^{(1+\alpha)}u = \frac{1}{x^2}D_x\left[x^4\left(u_x + u + u^2\right)\right]. \tag{2.85}$$

Finally, we make in the Eq. (2.81) the nonlocal change of the dependent variable by setting $f = J_t^{(1-\alpha)}u$. Since $D_t^{1-\alpha}J_t^{1-\alpha}u = u$, this equation can be rewritten as

$$D_t^{\alpha}u = \frac{1}{x^2}D_x\left[x^4\left(u_x + u + u^2\right)\right]. \tag{2.86}$$

We can formally rewrite the Eqs. (2.82)-(2.86) as

$$F(t,x,u,\mathcal{D}_t^{\gamma(\alpha)}u,u_x,u_{xx}) = \mathcal{D}_t^{\gamma(\alpha)}u - x^2D_xh(u,u_x) - 4xh(u,u_x) = 0. \tag{2.87}$$

Here we denote by $\mathcal{D}_t^{\gamma(\alpha)}$ any time-fractional differential operator in Eqs. (2.82)-(2.86), and the function h is

$$h(u,u_x) = u_x + u + u^2. \tag{2.88}$$

Every term in the function $h(u,u_x)$ describes a certain physical effect.

Ibragimov in [97] demonstrated that the only vector field generator of the Kompaneets Eq. (2.77) is the time-translations

$$V = \frac{\partial}{\partial t}.$$

The time-fractional generalizations of the Kompaneets equation do not admit the translation in time. The calculation shows that the Eqs. (2.82)-(2.86) have no Lie point symmetries. Nevertheless, the physically relevant approximations of these equations have nontrivial Lie point symmetries.

Lemma 2.9.1 *[64] Let the Eq. (2.87) be a diffusion-type time-fractional equation, i.e., the function h has the term u_x. Then the following approximations of this equation have the nontrivial Lie point symmetries:*

- *the Eq. (2.87) with $h = u_x$ admits*

$$V_1 = u\frac{\partial}{\partial u}, \qquad V_2 = x\frac{\partial}{\partial x}, \qquad V_g = g(t,x)\frac{\partial}{\partial u}; \tag{2.89}$$

- *the Eq. (2.87) with $h = u_x + u$ admits*

$$V_1 = u\frac{\partial}{\partial u}, \qquad V_g = g(t,x)\frac{\partial}{\partial u}; \tag{2.90}$$

- *the Eq. (2.87) with* $h = u_x + u^2$ *admits*

$$V_3 = V_2 - V_1 = x\frac{\partial}{\partial x} - u\frac{\partial}{\partial u}, \tag{2.91}$$

where the function $g(t, x)$ *is a solution of the equation*

$$\mathcal{D}_t^{\gamma(\alpha)} g = x^2 D_x h(g, g_x) + 4xh(g, g_x).$$

The above symmetries are valid for all considered types of the time-fractional derivatives $D_t^{\gamma(\alpha)} u$.

We can construct invariant solutions only for two types of approximations of the time-fractional Kompaneets equations that correspond to the following types of the function $h = u_x$ and $h = u_x + u^2$.

- $h = u_x$.

Then we can find the invariant solutions of Eq. (2.87) using a linear combination of the operators V_1 and V_2:

$$V = \beta V_1 + V_2 = \beta u\frac{\partial}{\partial u} + x\frac{\partial}{\partial x}.$$

The corresponding invariant solution has the form

$$u(t, x) = x^{\beta}(t).$$

Substituting this expression in the Eq. (2.87), we obtain the ordinary fractional differential equation for the function $z(t)$:

$$\mathcal{D}_t^{\gamma(\alpha)} z = \beta(\beta + 3)z. \tag{2.92}$$

For $\beta = 0$ or $\beta = -3$ this equation takes a very simple form

$$\mathcal{D}_t^{\gamma(\alpha)} z = 0. \tag{2.93}$$

For different time-fractional differential operators, the Eq. (2.93) has the following common solutions:

$$\begin{aligned}
&\mathcal{D}_t^{\gamma(\alpha)} = D_t^{\alpha} D_t : \quad z(t) = c_1 t^{\alpha} + c_2;\\
&\mathcal{D}_t^{\gamma(\alpha)} = {}^{C}D_t^{\alpha} : \quad z(t) = c_1;\\
&\mathcal{D}_t^{\gamma(\alpha)} = D_t^{1+\alpha} : \quad z(t) = c_1 t^{\alpha} + c_2 t^{\alpha-1};\\
&\mathcal{D}_t^{\gamma(\alpha)} = D_t^{\alpha} : \quad z(t) = c_1 t^{\alpha-1},
\end{aligned} \tag{2.94}$$

where c_1 and c_2 are arbitrary constants. Then, we get the following invariant solutions:

• for the approximation of the Eq. (2.82)

$$u = c_1 t^\alpha + c_2, \quad u = x^{-3}(c_1 t^\alpha + c_2);$$

• for the approximation of the Eq. (2.84)

$$u = c_1, \quad u = x^{-3} c_1;$$

• for the approximation of the Eq. (2.85)

$$u = t^{\alpha-1}(c_1 t + c_2), \quad u = x^{-3} t^{\alpha-1}(c_1 t + c_2)$$

• for the approximation of the Eq. (2.86)

$$u = c_1 t^{\alpha-1}, \quad u = c_1 x^{-3} t^{\alpha-1}.$$

Remark 2.9.1 *To give a physical interpretation of the obtained solutions, it is necessary to make the inverse change of variables $u \to f$ in accordance with the definitions of u for different generalizations of the Kompaneets equation. It is easy to show that after such inverse change of variables, for all obtained solutions we have $f(t) = 0$, i.e., these solutions are the stationary solutions of the corresponding approximations of the Eqs. (2.78)-(2.81).*
Now we consider the Eq. (2.92) with $\beta \neq 0$, $\beta \neq -3$, and denote $\lambda = \beta(\beta+3)$. The common solutions of the Eq. (2.92) for different time-fractional differential operators can be obtained using the Laplace transform. As a result, we arrive at the following invariant solutions:

• *For the approximation of the Eq. (2.82)*

$$u = x^\beta (c_1 E_{1+\alpha,1}(\lambda t^{1+\alpha}) + c_2 t^\alpha E_{1+\alpha,1+\alpha}(\lambda t^{1+\alpha})).$$

• *For the approximation of the Eq. (2.84)*

$$u = c_1 x^\beta E_{\alpha,1}(\lambda t^\alpha).$$

• *For the approximation of the Eq. (2.85)*

$$u = x^\beta t^{\alpha-1}(c_1 E_{1+\alpha,\alpha}(\lambda t^{\alpha+1}) + c_2 t E_{\alpha+1,\alpha+1}(\lambda t^\alpha)).$$

• *For the approximation of the Eq. (2.86)*

$$u = c_1 x^\beta t^{\alpha-1} E_{\alpha,\alpha}(\lambda t^\alpha).$$

Remark 2.9.2 *To give a physical interpretation, we note that after inverse change of variables $u \to f$ in these solutions, one can conclude that these solutions describe the dynamic regimes in which the width of the corresponding diffusion packets increases according to the power law.*

Now let us consider another case when $h = u_x + u^2$. The invariant solution corresponding to the operator

$$V = x\frac{\partial}{\partial x} - u\frac{\partial}{\partial u}$$

has the form

$$u = x^{-1}z(t).$$

Substituting it into the Eq. (2.87), we get the following ordinary fractional differential equation for the function $f(t)$:

$$\mathcal{D}_t^{\gamma(\alpha)} z = 2(z^2 - z). \tag{2.95}$$

We cannot present here any exact solution of this nonlinear equation for any time-fractional differential operators considered in this paper. Note that the Eq. (2.95) does not have Lie point symmetries.

2.10 Lie symmetry analysis of the time-fractional variant Boussinesq and coupled Boussinesq-Burger's equations

A completely integrable variant of the Boussinesq (VB) system is considered by Sachs [165, 62] as:

$$\begin{cases} u_t + v_x + uu_x = 0, \\ v_t + (uv)_x + u_{xxx} = 0. \end{cases} \tag{2.96}$$

This system models the water waves; u is velocity and v is the total depth. So physical solutions ought to have $v > 0$.
Another interesting model which arises in the study of fluids flow in a dynamic system, is the Boussinesq-Burger's (BB) equation [116]. Moreover, this equation describes the propagation of shallow water waves and can be written as:

$$\begin{cases} u_t - \frac{1}{2}v_x + 2uu_x = 0, \\ v_t - \frac{1}{2}u_{xxx} + 2(uv)_x = 0. \end{cases} \tag{2.97}$$

In general, it is difficult to find exact solutions of differential equations with fractional derivatives and we remember that investigation of some properties of fractional derivatives are much harder than the classical ones. Therefore, there is a strong motivation to find the exact solutions and Lie symmetries of famous equations like the Eqs. (2.96)-(2.97).
The time-fractional versions of Eqs. (2.96) and (2.97) have the following forms:

$$\begin{cases} \partial_t^\alpha u + v_x + uu_x = 0, \\ \partial_t^\alpha v + (uv)_x + u_{xxx} = 0, \end{cases} \tag{2.98}$$

and

$$\begin{cases} \partial_t^\alpha u - \frac{1}{2}v_x + 2uu_x = 0, \\ \partial_t^\alpha v - \frac{1}{2}u_{xxx} + 2(uv)_x = 0, \end{cases} \tag{2.99}$$

respectively, where $\partial_t^\alpha u := {}^{RL}D_t^\alpha u$ stands for Riemann-Liouville derivative.

Let us consider the Lie symmetry analysis of system of FPDEs [71, 67, 77, 176]:

$$\begin{cases} \Pi_1 := \partial_t^\alpha u - F_1(x,t,u,v,u_x,v_x,\ldots) = 0, \\ \Pi_2 := \partial_t^\alpha v - F_2(x,t,u,v,u_x,v_x,\ldots) = 0, \end{cases} \quad 0 < \alpha < 1. \tag{2.100}$$

The infinitesimal operator of the local point transformations Lie group which are admitted by Eq. (2.100) is:

$$X = \tau(x,t,u,v)\frac{\partial}{\partial t} + \xi(x,t,u,v)\frac{\partial}{\partial x} + {}^u\phi(x,t,u,v)\frac{\partial}{\partial u} + {}^v\phi(x,t,u,v)\frac{\partial}{\partial v}.$$

Applying the Lie theorem to an invariance condition yields:

$$Pr^{(\alpha,3)}X(\Pi_1 \wedge \Pi_2)|_{\Pi_1=0,\ \Pi_2=0} = 0, \tag{2.101}$$

where

$$\begin{aligned} Pr^{(\alpha,3)}X = X &+ {}^u\phi_\alpha^0\frac{\partial}{\partial_t^\alpha u} + {}^v\phi_\alpha^0\frac{\partial}{\partial_t^\alpha v} + {}^u\phi^x\frac{\partial}{\partial u_x} + {}^v\phi^x\frac{\partial}{\partial v_x} \\ &+ {}^u\phi^{xx}\frac{\partial}{\partial u_{xx}} + {}^u\phi^{xxx}\frac{\partial}{\partial u_{xxx}}, \end{aligned}$$

and

$$\begin{aligned} {}^u\phi^x &= D_x({}^u\phi) - u_xD_x(\xi) - u_tD_x(\tau), \\ {}^v\phi^x &= D_x({}^v\phi) - v_xD_x(\xi) - v_tD_x(\tau), \\ {}^u\phi^{xx} &= D_x({}^u\phi^x) - u_{xt}D_x(\tau) - u_{xx}D_x(\xi), \\ {}^u\phi^{xxx} &= D_x({}^u\phi^{xx}) - u_{xxt}D_x(\tau) - u_{xxx}D_x(\xi), \\ {}^u\phi_\alpha^0 &= D_t^\alpha({}^u\phi) + \xi D_t^\alpha(u_x) - D_t^\alpha(\xi u_x) + D_t^\alpha(D_t(\tau)u) - D_t^{\alpha+1}(\tau u) + \tau D_t^{\alpha+1}(u), \\ {}^v\phi_\alpha^0 &= D_t^\alpha({}^v\phi) + \xi D_t^\alpha(v_x) - D_t^\alpha(\xi v_x) + D_t^\alpha(D_t(\tau)v) - D_t^{\alpha+1}(\tau v) + \tau D_t^{\alpha+1}(v). \end{aligned}$$

Since the lower limit of integral in Riemann-Liouville fractional derivative is a unchanging value, so it should be invariant with respect to the point transformations:

$$\tau(x,t,u,v)|_{t=0} = 0. \tag{2.102}$$

The αth prolonged infinitesimal has the following form:

$${}^u\phi_\alpha^0 = D_t^\alpha({}^u\phi) + \xi D_t^\alpha(u_x) - D_t^\alpha(\xi u_x) + D_t^\alpha(D_t(\tau)u) - D_t^{\alpha+1}(\tau u) + \tau D_t^{\alpha+1}(u), \tag{2.103}$$

where D_t^α denotes the total fractional derivative. Let us remember the Leibniz rule in the fractional case:

$$D_t^\alpha[f(t)g(t)] = \sum_{n=0}^{\infty} \begin{pmatrix} \alpha \\ n \end{pmatrix} D_t^{\alpha-n} f(t) D_t^n g(t),$$

where

$$\begin{pmatrix} \alpha \\ n \end{pmatrix} = \frac{(-1)^{n-1}\alpha\Gamma(n-\alpha)}{\Gamma(1-\alpha)\Gamma(n+1)}.$$

Thus from (2.10) we can rewrite (2.103) as follows

$$ {}^u\phi_\alpha^0 = D_t^\alpha({}^u\phi) - \alpha D_t(\tau)\frac{\partial^\alpha u}{\partial t^\alpha} - \sum_{n=1}^{\infty} \begin{pmatrix} \alpha \\ n \end{pmatrix} D_t^n(\xi) D_t^{\alpha-n}(u_x)$$
$$ - \sum_{n=1}^{\infty} \begin{pmatrix} \alpha \\ n+1 \end{pmatrix} D_t^{n+1}(\tau) D_t^{\alpha-n}(u). \tag{2.104}$$

Also from the chain rule and setting $f(t) = 1$, we get

$$D_t^\alpha({}^u\phi) = \frac{\partial^\alpha({}^u\phi)}{\partial t^\alpha} + \left({}^u\phi_u \frac{\partial^\alpha u}{\partial t^\alpha} - u\frac{\partial^\alpha({}^u\phi_u)}{\partial t^\alpha}\right) + \left({}^u\phi_v \frac{\partial^\alpha v}{\partial t^\alpha} - v\frac{\partial^\alpha({}^u\phi_v)}{\partial t^\alpha}\right)$$
$$+ \sum_{n=1}^{\infty} \begin{pmatrix} \alpha \\ n \end{pmatrix} \frac{\partial^n({}^u\phi_u)}{\partial t^n} D_t^{\alpha-n}(u) + \sum_{n=1}^{\infty} \begin{pmatrix} \alpha \\ n \end{pmatrix} \frac{\partial^n({}^u\phi_v)}{\partial t^n} D_t^{\alpha-n}(v) + \chi_1 + \chi_2, \tag{2.105}$$

where

$$\chi_1 = \sum_{n=2}^{\infty}\sum_{m=2}^{n}\sum_{k=2}^{m}\sum_{r=0}^{k-1} \begin{pmatrix} \alpha \\ n \end{pmatrix} \begin{pmatrix} n \\ m \end{pmatrix} \begin{pmatrix} k \\ r \end{pmatrix} \frac{1}{k!}$$
$$\times \frac{t^{n-\alpha}}{\Gamma(n+1-\alpha)}[-u]^r \frac{\partial^m}{\partial t^m}[u^{k-r}] \frac{\partial^{n-m+k}({}^u\phi)}{\partial t^{n-m}\partial u^k},$$

and

$$\chi_2 = \sum_{n=2}^{\infty}\sum_{m=2}^{n}\sum_{k=2}^{m}\sum_{r=0}^{k-1} \begin{pmatrix} \alpha \\ n \end{pmatrix} \begin{pmatrix} n \\ m \end{pmatrix} \begin{pmatrix} k \\ r \end{pmatrix} \frac{1}{k!}$$
$$\times \frac{t^{n-\alpha}}{\Gamma(n+1-\alpha)}[-v]^r \frac{\partial^m}{\partial t^m}[v^{k-r}] \frac{\partial^{n-m+k}({}^u\phi)}{\partial t^{n-m}\partial v^k}.$$

Therefore

$$
\begin{aligned}
{}^u\phi_\alpha^0 =& \frac{\partial^\alpha({}^u\phi)}{\partial t^\alpha} + ({}^u\phi_u - \alpha D_t(\tau))\frac{\partial^\alpha u}{\partial t^\alpha} - u\frac{\partial^\alpha({}^u\phi_u)}{\partial t^\alpha} + \chi_1 + \chi_2 \\
&+ \left({}^u\phi_v\frac{\partial^\alpha v}{\partial t^\alpha} - v\frac{\partial^\alpha({}^u\phi_v)}{\partial t^\alpha}\right) + \sum_{n=1}^{\infty}\left(\begin{array}{c}\alpha\\ n\end{array}\right)\frac{\partial^n({}^u\phi_v)}{\partial t^n}D_t^{\alpha-n}(v) \\
&+ \sum_{n=1}^{\infty}\left[\left(\begin{array}{c}\alpha\\ n\end{array}\right)\frac{\partial^n({}^u\phi_u)}{\partial t^n} - \left(\begin{array}{c}\alpha\\ n+1\end{array}\right)D_t^{n+1}(\tau)\right]D_t^{\alpha-n}(u) \\
&- \sum_{n=1}^{\infty}\left(\begin{array}{c}\alpha\\ n\end{array}\right)D_t^n(\xi)D_t^{\alpha-n}(u_x).
\end{aligned}
$$

Similarly, we can obtain

$$
\begin{aligned}
{}^v\phi_\alpha^0 =& \frac{\partial^\alpha({}^v\phi)}{\partial t^\alpha} + ({}^v\phi_v - \alpha D_t(\tau))\frac{\partial^\alpha v}{\partial t^\alpha} - v\frac{\partial^\alpha({}^v\phi_v)}{\partial t^\alpha} + \chi_3 + \chi_4 \\
&+ \left({}^v\phi_u\frac{\partial^\alpha u}{\partial t^\alpha} - u\frac{\partial^\alpha({}^v\phi_u)}{\partial t^\alpha}\right) + \sum_{n=1}^{\infty}\left(\begin{array}{c}\alpha\\ n\end{array}\right)\frac{\partial^n({}^v\phi_u)}{\partial t^n}D_t^{\alpha-n}(u) \\
&+ \sum_{n=1}^{\infty}\left[\left(\begin{array}{c}\alpha\\ n\end{array}\right)\frac{\partial^n({}^v\phi_v)}{\partial t^n} - \left(\begin{array}{c}\alpha\\ n+1\end{array}\right)D_t^{n+1}(\tau)\right]D_t^{\alpha-n}(v) \\
&- \sum_{n=1}^{\infty}\left(\begin{array}{c}\alpha\\ n\end{array}\right)D_t^n(\xi)D_t^{\alpha-n}(v_x),
\end{aligned}
$$

where

$$
\begin{aligned}
\chi_3 =& \sum_{n=2}^{\infty}\sum_{m=2}^{n}\sum_{k=2}^{m}\sum_{r=0}^{k-1}\left(\begin{array}{c}\alpha\\ n\end{array}\right)\left(\begin{array}{c}n\\ m\end{array}\right)\left(\begin{array}{c}k\\ r\end{array}\right)\frac{1}{k!} \\
&\times \frac{t^{n-\alpha}}{\Gamma(n+1-\alpha)}[-u]^r\frac{\partial^m}{\partial t^m}[u^{k-r}]\frac{\partial^{n-m+k}({}^v\phi)}{\partial t^{n-m}\partial u^k},
\end{aligned}
$$

and

$$
\begin{aligned}
\chi_4 =& \sum_{n=2}^{\infty}\sum_{m=2}^{n}\sum_{k=2}^{m}\sum_{r=0}^{k-1}\left(\begin{array}{c}\alpha\\ n\end{array}\right)\left(\begin{array}{c}n\\ m\end{array}\right)\left(\begin{array}{c}k\\ r\end{array}\right)\frac{1}{k!} \\
&\times \frac{t^{n-\alpha}}{\Gamma(n+1-\alpha)}[-v]^r\frac{\partial^m}{\partial t^m}[v^{k-r}]\frac{\partial^{n-m+k}({}^v\phi)}{\partial t^{n-m}\partial v^k}.
\end{aligned}
$$

Now, we utilize the developed Lie symmetry method to the Eqs. (2.98)-(2.99) and then we obtain the corresponding reductions of these equations. According to the Lie theory, determining equations of the time-fractional VB equation (2.98) can be extracted from applying $Pr^{(\alpha,3)}X$ to Eq. (2.98) as follows:

$$\begin{aligned}
&\tau_u = \tau_v = \tau_x = \xi_t = \xi_u = \xi_v = {}^u\phi_v = {}^u\phi_{uu} = {}^v\phi_{vv} = 0,\\
&{}^u\phi_{xu} - \xi_{xx} = 0,\\
&-{}^v\phi_v - 3\xi_x + \alpha\tau_t + {}^u\phi_u = 0,\\
&{}^v\phi_u - u\xi_x + \alpha u\tau_t + {}^u\phi_v = 0,\\
&(1-\alpha)\tau_{tt} + 2\,{}^v\phi_{tv} = 0, \quad (1-\alpha)\tau_{tt} + 2\,{}^u\phi_{tu} = 0,\\
&(2-\alpha)\tau_{ttt} + 3\,{}^v\phi_{ttv} = 0, \quad (2-\alpha)\tau_{ttt} + 3\,{}^u\phi_{ttu} = 0,\\
&\alpha\tau_t - \xi_x - {}^u\phi_u + {}^v\phi_v = 0,\\
&\alpha u\tau_t - u\xi_x + v\,{}^u\phi_v + 3\,{}^u\phi_{xxv} + {}^u\phi = 0,\\
&\alpha v\tau_t - v\xi_x - \xi_{xxx} - v\,{}^v\phi_v + 3\,{}^u\phi_{xxu} + {}^v\phi + v\,{}^u\phi_u + u\,{}^v\phi_u = 0,\\
&u\,{}^u\phi_x + {}^v\phi_x - u\partial_t^\alpha({}^u\phi_u) + \partial_t^\alpha({}^u\phi) = 0,\\
&v\,{}^u\phi_x + u\,{}^v\phi_x - v\partial_t^\alpha({}^v\phi_v) + \partial_t^\alpha({}^v\phi) + {}^u\phi_{xxx} = 0,\\
&\sum_{k=3}^{\infty}\binom{\alpha}{k}\frac{\partial^{k+1}}{\partial t^k\partial u}({}^u\phi)\times D_{t^{\alpha-k}}u - \sum_{k=3}^{\infty}\frac{\binom{\alpha}{k}}{1+k}\big[(\alpha-k)\,D_{t^{\alpha-k}}u\times D_{t^{1+k}}\tau\\
&+ (k+1)D_{t^{\alpha-k}}u_x\times D_{t^k}\xi\big] = 0,\\
&\sum_{k=3}^{\infty}\binom{\alpha}{k}\frac{\partial^{k+1}}{\partial t^k\partial v}({}^v\phi)\times D_{t^{\alpha-k}}v - \sum_{k=3}^{\infty}\frac{\binom{\alpha}{k}}{1+k}\big[(\alpha-k)\,D_{t^{\alpha-k}}v\times D_{t^{1+k}}\tau\\
&+ (k+1)D_{t^{\alpha-k}}v_x\times D_{t^k}\xi\big] = 0.
\end{aligned} \tag{2.106}$$

Solving Eqs. (2.106) and applying the condition (2.102), we obtain the following infinitesimals:

$$\tau = 2tc_2, \quad \xi = c_1 + \alpha x c_2, \quad {}^u\phi = -\alpha u c_2, \quad {}^v\phi = -2\alpha v c_2, \tag{2.107}$$

where c_1 and c_2 are arbitrary constants and therefore:

$$X_1 = \frac{\partial}{\partial x}, \quad X_2 = 2t\frac{\partial}{\partial t} + \alpha x\frac{\partial}{\partial x} - \alpha u\frac{\partial}{\partial u} - 2\alpha v\frac{\partial}{\partial v}. \tag{2.108}$$

Let us consider the reduction of Eq. (2.98) corresponding to the second vector field X_2. The similarity variables and similarity transformations corresponding to X_2 take the following form:

$$\begin{cases} u(x,t) = t^{-\frac{\alpha}{2}}\mathcal{F}(\zeta), \\ v(x,t) = t^{-\alpha}\mathcal{G}(\zeta), \end{cases} \quad \zeta = xt^{-\frac{\alpha}{2}}. \tag{2.109}$$

Theorem 18 *Using (2.112), the time-fractional VB equation (2.98) is reduced to the following nonlinear system of FODEs:*

$$\begin{cases} \left(\mathcal{P}_{\frac{2}{\alpha}}^{1-\frac{3\alpha}{2},\alpha}\mathcal{F}\right)(\zeta) + \mathcal{G}'(\zeta) + \mathcal{F}(\zeta)\mathcal{F}'(\zeta) = 0, \\ \left(\mathcal{P}_{\frac{2}{\alpha}}^{1-\alpha,\alpha}\mathcal{G}\right)(\zeta) + \mathcal{F}(\zeta)\mathcal{G}'(\zeta) + \mathcal{F}'(\zeta)\mathcal{G}(\zeta) + \mathcal{F}'''(\zeta) = 0. \end{cases} \tag{2.110}$$

Proof: From the definition of Riemann-Liouville fractional derivative and similarity variables related to X_2, we have

$$\frac{\partial^\alpha u}{\partial t^\alpha} = \frac{\partial^n}{\partial t^n}\Big[\frac{1}{\Gamma(n-\alpha)}\int_0^t (t-s)^{n-\alpha-1} s^{-\frac{\alpha}{2}}\mathcal{F}\left(xs^{\frac{-\alpha}{2}}\right) ds\Big]. \tag{2.111}$$

If we take the change of variable $\rho = \frac{t}{s}$, then $ds = -\frac{t}{\rho^2}d\rho$, so from (2.111) we have

$$\frac{\partial^\alpha u}{\partial t^\alpha} = \frac{\partial^n}{\partial t^n}\Big[t^{n-\frac{3\alpha}{2}}\left(\mathcal{K}_{\frac{2}{\alpha}}^{1-\frac{\alpha}{2},n-\alpha}\mathcal{F}\right)(\zeta)\Big].$$

Moreover $\zeta = xt^{\frac{-\alpha}{2}}$ concludes

$$t\frac{\partial}{\partial t}\phi(\zeta) = t\frac{\partial\zeta}{\partial t}\frac{d\phi(\zeta)}{d\zeta} = -\frac{\alpha}{2}\zeta\frac{d\phi(\zeta)}{d\zeta}.$$

Therefore by setting $\Phi(\zeta) = \left(\mathcal{K}_{\frac{2}{\alpha}}^{1-\frac{\alpha}{2},n-\alpha}\mathcal{F}\right)(\zeta)$ one can get

$$\begin{aligned}
\frac{\partial^n}{\partial t^n}\Big[t^{n-\frac{3\alpha}{2}}\Phi(\zeta)\Big] &= \frac{\partial^{n-1}}{\partial t^{n-1}}\Big[\frac{\partial}{\partial t}\left(t^{n-\frac{3\alpha}{2}}\Phi(\zeta)\right)\Big]\\
&= \frac{\partial^{n-1}}{\partial t^{n-1}}\Big[t^{n-\frac{3\alpha}{2}-1}\left(n-\frac{3\alpha}{2}-\frac{\alpha}{2}\zeta\frac{d}{d\zeta}\right)\Phi(\zeta)\Big]\\
&= \cdots = t^{-\frac{3\alpha}{2}}\prod_{j=0}^{n-1}\left(1-\frac{3\alpha}{2}+j-\frac{\alpha}{2}\zeta\frac{d}{d\zeta}\right)\Phi(\zeta) = t^{-\frac{3\alpha}{2}}\left(\mathcal{P}_{\frac{2}{\alpha}}^{1-\frac{3\alpha}{2},\alpha}\mathcal{F}\right)(\zeta).
\end{aligned}$$

Therefore

$$\frac{\partial^\alpha u}{\partial t^\alpha} = t^{-\frac{3\alpha}{2}}\left(\mathcal{P}_{\frac{2}{\alpha}}^{1-\frac{3\alpha}{2},\alpha}\mathcal{F}\right)(\zeta).$$

In a similar way, we can obtain

$$\frac{\partial^\alpha v}{\partial t^\alpha} = t^{-2\alpha}\left(\mathcal{P}_{\frac{2}{\alpha}}^{1-\alpha,\alpha}\mathcal{G}\right)(\zeta),$$

which completes the proof. □

Now, similarly to the time-fractional VB equation, by applying $Pr^{(\alpha,3)}X$ to Eq. (2.99), we can find a system of an overdetermined system of PDEs and FPDEs, from which by solving them we obtain the following infinitesimals:

$$\tau = 2tc_2,\quad \xi = c_1 + \alpha xc_2,\quad {}^u\phi = -\alpha uc_2,\quad {}^v\phi = -2\alpha vc_2,$$

where c_1 and c_2 are arbitrary constants and therefore:

$$X_1 = \frac{\partial}{\partial x},\quad X_2 = 2t\frac{\partial}{\partial t} + \alpha x\frac{\partial}{\partial x} - \alpha u\frac{\partial}{\partial u} - 2\alpha v\frac{\partial}{\partial v}.$$

Let us consider the reduction of Eq. (2.99) corresponding to the second vector field X_2. The similarity variables and similarity transformations corresponding to X_2 take the following form:

$$\begin{cases} u(x,t) = t^{-\frac{\alpha}{2}}\mathcal{F}(\zeta), \\ v(x,t) = t^{-\alpha}\mathcal{G}(\zeta), \end{cases}\quad \zeta = xt^{-\frac{\alpha}{2}}. \tag{2.112}$$

Theorem 19 *Using (2.112), the time-fractional BB equation (2.99) is reduced to the following nonlinear system of FODEs:*

$$\begin{cases} \left(\mathcal{P}_{\frac{2}{\alpha}}^{1-\frac{3\alpha}{2},\alpha}\mathcal{F}\right)(\zeta) - \frac{1}{2}\mathcal{G}'(\zeta) + 2\mathcal{F}(\zeta)\mathcal{F}'(\zeta) = 0, \\ \left(\mathcal{P}_{\frac{2}{\alpha}}^{1-\alpha,\alpha}\mathcal{G}\right)(\zeta) + 2\mathcal{F}(\zeta)\mathcal{G}'(\zeta) + 2\mathcal{F}'(\zeta)\mathcal{G}(\zeta) - \frac{1}{2}\mathcal{F}'''(\zeta) = 0. \end{cases} \tag{2.113}$$

Proof: Similar to proof of Theorem 18 □

2.10.1 Exact solutions of time-fractional VB and BB equations by invariant subspace method

In this section, we briefly describe the invariant subspace method applicable to the coupled time FPDEs of the form:

$$\begin{cases} \partial_t^\alpha u = \Xi_1[u,v], \\ \partial_t^\alpha v = \Xi_2[u,v], \end{cases} \quad \alpha \in \mathbb{R}_+, \tag{2.114}$$

where $\Xi_1[u,v]$ and $\Xi_2[u,v]$ are differential operators of u and v with respect to the independent variable x.

Definition 16 *Finite dimensional linear spaces*

$$\mathcal{W}_{n_q}^q = span\{\omega_1^q(x), \omega_2^q(x), \cdots, \omega_{n_q}^q(x)\}, \ q = 1, 2,$$

are said to be invariant subspaces with respect to the system of (2.114), if $\Xi_1[\mathcal{W}_{n_1}^1 \times \mathcal{W}_{n_2}^2] \subseteq \mathcal{W}_{n_1}^1 \times \mathcal{W}_{n_2}^2$ *and* $\Xi_2[\mathcal{W}_{n_1}^1 \times \mathcal{W}_{n_2}^2] \subseteq \mathcal{W}_{n_1}^1 \times \mathcal{W}_{n_2}^2$.

Suppose that the Eqs. (2.114) admit an invariant subspace $\mathcal{W}_{n_1}^1 \times \mathcal{W}_{n_2}^2$. Then from the above definition, there exist the expansion coefficient functions $\psi_1^1, \psi_2^1, \cdots, \psi_{n_1}^1$ and $\psi_1^2, \psi_2^2, \cdots, \psi_{n_2}^2$ such that

$$\Xi_q\left[\sum_{i=1}^{n_1} \lambda_i^1 \omega_i^1(x), \sum_{i=1}^{n_2} \lambda_i^2 \omega_i^2(x)\right] = \sum_{i=1}^{n_q} \psi_i^q(\lambda_1^1, \lambda_2^1, \cdots, \lambda_{n_1}^1, \lambda_1^2, \lambda_2^2, \cdots, \lambda_{n_2}^2)\omega_i^q(x),$$
$$\lambda_i^q \in \mathbb{R}, \ i = 1, \cdots, n_q, \ q = 1, 2. \tag{2.115}$$

Hence

$$\begin{cases} u(x,t) = \sum_{i=1}^{n_1} \lambda_i^1(t)\omega_i^1(x), \\ v(x,t) = \sum_{i=1}^{n_2} \lambda_i^2(t)\omega_i^2(x), \end{cases} \tag{2.116}$$

is the solution of Eq. (2.114), if the expansion coefficients $\lambda_i^q(t), \ q = 1, 2$, satisfy a system of FODEs:

$$\partial_t^\alpha \lambda_i^q(t) = \psi_i^q(\lambda_1^1(t), \lambda_2^1(t), \ldots, \lambda_{n_1}^1(t), \lambda_1^2(t), \lambda_2^2(t), \ldots, \lambda_{n_2}^2(t)),$$
$$i = 1, \ldots, n_q, \ q = 1, 2. \tag{2.117}$$

Now, in order to find the invariant subspace $\mathcal{W}_{n_1}^1 \times \mathcal{W}_{n_2}^2$ of a given fractional equation, one can use the following theorem [60].

Theorem 20 *Let functions $\omega_1^q(x), \ldots, \omega_{n_q}^q(x)$, $q = 1, 2$ form the fundamental set of solutions of a linear n_q-th order ODEs*

$$\mathcal{L}_q[y_q] \equiv y_q^{(n_q)} + a_1^q(x) y_q^{(n_q-1)} + \cdots + a_{n_q-1}^q(x) y_q' + a_n^q(x) y_q = 0, \ q = 1, 2 \tag{2.118}$$

and let Ξ_1, Ξ_2 be smooth enough functions. Then the subspace $\mathcal{W}_{n_1}^1 \times \mathcal{W}_{n_2}^2$ is invariant with respect to the operators Ξ_1 and Ξ_2 if and only if

$$\mathcal{L}_q(\Xi_q[u, v])|_{\mathcal{L}_1[u]=0 \wedge \mathcal{L}_2[v]=0} = 0, \ q = 1, 2. \tag{2.119}$$

Let us assume

$$\begin{cases} \dfrac{\partial^\alpha u}{\partial t^\alpha} = \Xi_1[u, v] = -\dfrac{\partial v}{\partial x} - u\dfrac{\partial u}{\partial x}, \\ \dfrac{\partial^\alpha v}{\partial t^\alpha} = \Xi_2[u, v] = -\dfrac{\partial v}{\partial x} u - \dfrac{\partial u}{\partial x} v - \dfrac{\partial^3 u}{\partial x^3}, \end{cases} \quad \alpha \in (0, 1). \tag{2.120}$$

In order to find the invariant subspace, we consider $n_1 = 2$, $n_2 = 2$. In this case, the invariant subspace $\mathcal{W}_2^1 \times \mathcal{W}_2^2$ is defined by two second order linear ODEs

$$\mathcal{W}_2^1 = \{y | \mathcal{L}_1[y] = y'' + a_1 y' + a_0 y = 0\},$$
$$\mathcal{W}_2^2 = \{z | \mathcal{L}_2[z] = z'' + b_1 z' + b_0 z = 0\},$$

where a_0, a_1, b_0 and b_1 are constants to be determined. The corresponding invariance conditions read as

$$(\mathcal{D}^2 \Xi_1 + a_1 \mathcal{D} \Xi_1 + a_0 \Xi_1)|_{(u,v) \in \mathcal{W}_2^1 \times \mathcal{W}_2^2} = 0, \tag{2.121}$$
$$(\mathcal{D}^2 \Xi_2 + b_1 \mathcal{D} \Xi_2 + b_0 \Xi_2)|_{(u,v) \in \mathcal{W}_2^1 \times \mathcal{W}_2^2} = 0, \tag{2.122}$$

where $\mathcal{D}$ is a differentiation operator with respect to x.
Eqs. (2.121) yields an overdetermined system of algebraic equations as follows:

$$\begin{aligned}
&2a_1 b_1 - b_1^2 - a_0 + b_0 = 0, \\
&2a_1 = 0, \\
&3a_0 = 0, \\
&a_1 b_0 - b_0 b_1 = 0, \\
&3a_0 = 0, \\
&3a_1 + b_1 = 0, \\
&-a_1^2 + a_1 b_1 + a_0 + 2b_0 = 0, \\
&-a_1^4 + a_1^3 b_1 + 3a_0 a_1^2 - 2a_0 a_1 b_1 - a_1^2 b_0 - a_0^2 + a_0 b_0 = 0, \\
&-a_0 a_1 + a_0 b_1 = 0, \\
&-a_0 a_1^3 + a_0 a_1^2 b_1 + 2a_0^2 a_1 - a_0^2 b_1 - a_0 a_1 b_0 = 0,
\end{aligned}$$

which can be solved in general and we obtain just the trivial solution $a_0 = a_1 = b_0 = b_1 = 0$. Therefore

$$\mathcal{L}_1[y] = y'' = 0, \ \ \mathcal{L}_2[z] = z'' = 0. \tag{2.123}$$

Thus, $\mathcal{W}_2^1 = span\{1, x\}$ and $\mathcal{W}_2^2 = span\{1, x\}$ and so

$$\mathcal{W}_2^1 \times \mathcal{W}_2^2 = span\left(\{1, x\} \times \{1, x\}\right). \tag{2.124}$$

This invariant subspace takes the exact solution of time-fractional VB equation as

$$\begin{cases} u(x,t) = \lambda_1^1(t) + \lambda_2^1(t)x, \\ v(x,t) = \lambda_1^2(t) + \lambda_2^2(t)x, \end{cases} \tag{2.125}$$

where $\lambda_1^1(t), \lambda_2^1(t), \lambda_1^2(t)$ and $\lambda_2^2(t)$ are unknown functions to be determined. Substituting (2.125) in to the system (2.120) we obtain the following system of FODEs:

$$\begin{cases} \dfrac{\partial^\alpha \lambda_1^1(t)}{\partial t^\alpha} = -\lambda_1^1(t)\lambda_2^1(t) - \lambda_2^2(t), \\ \dfrac{\partial^\alpha \lambda_2^1(t)}{\partial t^\alpha} = -(\lambda_2^1(t))^2, \\ \dfrac{\partial^\alpha \lambda_1^2(t)}{\partial t^\alpha} = -\lambda_1^1(t)\lambda_2^2(t) - \lambda_2^1(t)\lambda_1^2(t), \\ \dfrac{\partial^\alpha \lambda_2^2(t)}{\partial t^\alpha} = -2\lambda_2^1(t)\lambda_2^2(t). \end{cases} \tag{2.126}$$

Second and fourth equations of above system yield:

$$\lambda_2^1(t) = \frac{-\Gamma(1-\alpha)}{\Gamma(1-2\alpha)} t^{-\alpha},$$

and

$$\lambda_2^2(t) = 0.$$

Proceeding further, from the first and third equations, we find

$$\lambda_1^1(t) = \lambda_1^2(t) = t^{-\alpha}.$$

Thus, the obtained exact solution for time-fractional VB equation by invariant subspace method can be written as:

$$\begin{cases} u(x,t) = t^{-\alpha} - \dfrac{\Gamma(1-\alpha)}{\Gamma(1-2\alpha)} t^{-\alpha} x, \\ v(x,t) = t^{-\alpha}, \end{cases} \quad \alpha \in (0,1) - \{1/2\}. \tag{2.127}$$

Similarly, let us assume

$$\begin{cases} \frac{\partial^\alpha u}{\partial t^\alpha} = \Xi_1[u,v] = \frac{1}{2}\frac{\partial v}{\partial x} - 2u\frac{\partial u}{\partial x}, \\ \frac{\partial^\alpha v}{\partial t^\alpha} = \Xi_2[u,v] = -2\frac{\partial v}{\partial x}u - 2\frac{\partial u}{\partial x}v + \frac{1}{2}\frac{\partial^3 u}{\partial x^3}, \end{cases} \quad \alpha \in (0,1). \tag{2.128}$$

For this equation, we can find that the invariant subspace has the form:

$$\mathcal{W}_2^1 \times \mathcal{W}_2^2 = span\left(\{1, x\} \times \{1, x\}\right).$$

Thus, we look for an exact solution of the form

$$u(x,t) = \lambda_1^1(t) + \lambda_2^1(t)x, \qquad v(x,t) = \lambda_1^2(t) + \lambda_2^2(t)x,$$

where the coefficients are functions satisfying the following system of FODEs:

$$\begin{cases} \dfrac{\partial^\alpha \lambda_1^1(t)}{\partial t^\alpha} = \dfrac{1}{2}\lambda_2^2(t) - 2\lambda_1^1(t)\lambda_2^1(t), \\ \dfrac{\partial^\alpha \lambda_2^1(t)}{\partial t^\alpha} = -2(\lambda_2^1(t))^2, \\ \dfrac{\partial^\alpha \lambda_1^2(t)}{\partial t^\alpha} = -2\lambda_1^1(t)\lambda_2^2(t) - 2\lambda_2^1(t)\lambda_1^2(t), \\ \dfrac{\partial^\alpha \lambda_2^2(t)}{\partial t^\alpha} = -4\lambda_2^1(t)\lambda_2^2(t). \end{cases} \tag{2.129}$$

By solving the system of (2.129) we obtain the final exact solution of Eq. (2.99) in the following form:

$$\begin{cases} u(x,t) = t^{-\alpha} - \dfrac{1}{2}\dfrac{\Gamma(1-\alpha)}{\Gamma(1-2\alpha)}t^{-\alpha}x, \\ v(x,t) = t^{-\alpha}, \end{cases} \qquad \alpha \in (0,1) - \{1/2\}.$$

Chapter 3

Analytical Lie group approach for solving the fractional integro-differential equations

In addition to the differential equations, analysis of the symmetry groups for integro-differential equations were also investigated previously. The Lie-Bäcklund type operators are used to investigate the solutions of nonlocal determining equations in [30, 130, 99, 151]. Also, the Lie point groups are utilized by some authors [162, 151, 11, 188]. The symmetry group analysis of nonlocal differential equations is introduced by Özer [152] which is different from the Lie-Bäcklund type group of transformations.

Recently, the symmetry group analysis was developed for some special classes of fractional differential equations. The major difference between the developed Lie symmetry method and other ones is the prolongation formula, first derived by Gazizov et al. [66]. This method was applied for some time-fractional differential equations in [71, 77, 75]. Beside the Lie symmetry method, some other analytical and semi-analytical approaches are investigated for the solutions of fractional integro-differential equations (FIDEs) in literature [168, 3, 134, 182, 106].

In the current chapter, a new class of Lie symmetry methods is developed for dealing with the FIDEs. In our presented analytical method which is based upon the symmetry groups, the nonlocal variables of an equation are taken into account as independent variables. The determining equations of differential equations with nonlocal structure contain some extra nonlocal variables that differ from nonlocal variables of the original equation. A point symmetry group extension of the original equation is determined based on the nonlocal variables and this makes it difficult to solve the derived equations. Therefore, a new method is discussed for solving derived determining equations.

3.1 Lie groups of transformations for FIDEs

Consider a FIDE of the form [154]

$$\mathcal{D}_x^\alpha y(x) = F(x, y(x), \mathcal{T}(y)(x)), \tag{3.1}$$

with

$$\mathcal{T}(y)(x) = \int_\Omega f(x, t, y(t))dt, \tag{3.2}$$

where f and F are sufficiently smooth functions, Ω is a given interval of real line $\mathbb{R}$ and $x, y(x)$ are independent and dependent variables, respectively. In Eq. (3.1), $\mathcal{D}_x^\alpha := {}^{RL}D_x^\alpha$ denotes the fractional differential operator of order $\alpha > 0$ in the Riemann-Liouville sense. For the Eq. (3.1), consider the point transformations group

$$\begin{aligned} \tilde{x} &= \exp(\varepsilon V)(x) = x + \varepsilon\xi(x, y) + O(\varepsilon^2), \\ \tilde{y} &= \exp(\varepsilon V)(y) = y + \varepsilon\eta(x, y) + O(\varepsilon^2), \end{aligned} \tag{3.3}$$

with infinitesimal generator

$$V = \xi(x, y)\frac{\partial}{\partial x} + \eta(x, y)\frac{\partial}{\partial y}, \tag{3.4}$$

where the coefficients are sufficiently smooth functions. Moreover, a transformation for the integration variable t can be written as:

$$\tilde{t} = t + \varepsilon\xi(t, y(t)) + O(\varepsilon^2). \tag{3.5}$$

According to the Lie theory, construction of the symmetry group is equivalent to determine the infinitesimal transformations (3.3), i.e., $\xi(x, y)$ and $\eta(x, y)$.

Definition 17 *Eq. (3.1) is invariant under the group of transformations (3.3), if*

$$D_{\tilde{x}}^\alpha \tilde{y}(\tilde{x}) = F(\tilde{x}, \tilde{y}(\tilde{x}), \tilde{\mathcal{T}}(\tilde{y})(\tilde{x})), \quad \tilde{\mathcal{T}}(\tilde{y})(\tilde{x}) = \int_{\tilde{\Omega}} f(\tilde{x}, \tilde{t}, \tilde{y}(\tilde{t}))d\tilde{t}, \tag{3.6}$$

whenever Eq. (3.1) holds and $\tilde{\Omega}$ is the image of Ω under the transformation $t \to \tilde{t}$ given by (3.5). This group of transformations is called a point symmetry group for the Eq. (3.1).

3.2 The invariance criterion for FIDEs

In this section, we introduce a systematic way to obtain the invariant condition of Eq. (3.1) and corresponding solutions of a nonlocal determining equation. To do this, we first determine a prolongation on $\mathcal{T}(y)$ and $\mathcal{D}_x^{\alpha}y$ with infinitesimal generators (3.3). Moreover, differentiation of the integration variable t and its corresponding transformation can be calculated as

$$J(t) := (\tilde{t})' = 1 + \varepsilon D_t\xi(t, y(t)) + O(\varepsilon^2), \tag{3.7}$$

where D_t denotes the total derivative with respect to t. Then, by changing the variable of integration in (3.6), we obtain

$$\tilde{\mathcal{T}}(\tilde{y})(\tilde{x}) = \int_{\Omega} f(\tilde{x}, \tilde{t}, \tilde{y}(\tilde{t}))J(t)dt, \tag{3.8}$$

and using the Taylor expansion, the integrand in (3.6) can be represented as follows:

$$f(\tilde{x}, \tilde{t}, \tilde{y}(\tilde{t})) = f(x, t, y(t)) + \varepsilon(\mathcal{Q}_m f)(x, t, y(t)) + O(\varepsilon^2), \tag{3.9}$$

where

$$\mathcal{Q}_m = \xi(x, y(x))\frac{\partial}{\partial x} + \xi(t, y(t))\frac{\partial}{\partial t} + \eta(t, y(t))\frac{\partial}{\partial y(t)}. \tag{3.10}$$

Substituting Eqs. (3.7) and (3.9) into (3.8) yields

$$\tilde{\mathcal{T}}(\tilde{y})(\tilde{x}) = \mathcal{T}(y)(x) + \varepsilon P_{\mathcal{T}}(y)(x) + O(\varepsilon^2), \tag{3.11}$$

in which the nonlinear operator $P_{\mathcal{T}}$ is defined by

$$P_{\mathcal{T}}(y)(x) = \int_{\Omega} \left[\mathcal{Q}_m + D_t\xi(t, y(t))\right] f(x, t, y(t))dt. \tag{3.12}$$

The prolongation of point transformations group (3.3) on the nonlocal variable $A = \mathcal{T}(y)$ is defined by using (3.11). Therefore, we can consider the modified prolongation as:

$$Pr^{\mathcal{T}}V = V + P_{\mathcal{T}}(y)\frac{\partial}{\partial(\mathcal{T}(y))}, \tag{3.13}$$

where V is defined by (3.4).
By extending the transformations (3.3) to the Riemann-Liouville operator $\mathcal{D}_x^{\alpha}y$, we can determine the infinitesimal transformation of fractional derivatives as

$$D_{\tilde{x}}^{\alpha}\tilde{y}(\tilde{x}) = \mathcal{D}_x^{\alpha}y(x) + \varepsilon\eta^{\alpha} + O(\varepsilon^2), \tag{3.14}$$

where η^{α} is given by the prolongation formula:

$$\eta^{\alpha} = \mathcal{D}_x^{\alpha}\eta + \mathcal{D}_x^{\alpha}(\mathcal{D}_x(\xi)y) + \xi\mathcal{D}_x^{\alpha+1}y - \mathcal{D}_x^{\alpha+1}(\xi y). \tag{3.15}$$

Now, we are ready to introduce a new infinitesimal generator containing both integral and fractional terms as follows:

$$Pr^{\mathcal{T},\alpha}V = Pr^{\mathcal{T}}V + \eta^{\alpha}\frac{\partial}{\partial \mathcal{D}_x^{\alpha}y}. \tag{3.16}$$

According to the infinitesimal criterion, Eq. (3.1) admits the transformation group (3.3) iff the prolonged vector field $Pr^{\mathcal{T},\alpha}V$ annihilates (3.1) on its solution manifold, i.e.,

$$Pr^{\mathcal{T},\alpha}V(\Delta)\Big|_{\Delta=0} = 0, \qquad \Delta = \mathcal{D}_x^\alpha y - F(x, y, \mathcal{T}(y)). \tag{3.17}$$

In order to determine the vector fields (3.4) admitted by Eq.(3.1), it is necessary to obtain the concrete expression for $Pr^{\mathcal{T},\alpha}V$. Since we obtain the explicit expression for $Pr^{\mathcal{T}}V$ in (3.13), here we concentrate on the expression for η^α. Using the generalized Leibniz rule

$$\mathcal{D}_x^\alpha(f(x)g(x)) = \sum_{n=0}^{\infty}\binom{\alpha}{n}\mathcal{D}_x^{\alpha-n}f(x)\mathcal{D}_x^n g(x),$$

with

$$\binom{\alpha}{n} = \frac{(-1)^{n-1}\alpha\Gamma(n-\alpha)}{\Gamma(1-\alpha)\Gamma(n+1)}, \quad \mathcal{D}_x^0 f(x) = f(x), \quad \mathcal{D}_x^{n+1}f(x) = \mathcal{D}_x(\mathcal{D}_x^n f(x)),$$

and $\binom{\alpha}{0} = 1$, $\binom{\alpha+1}{1} = \alpha + 1$, we obtain

$$\begin{aligned}
\eta^\alpha &= \mathcal{D}_x^\alpha\eta + \sum_{n=0}^{\infty}\binom{\alpha}{n}\mathcal{D}_x^{\alpha-n}y\mathcal{D}_x^n(\mathcal{D}_x(\xi)) + \xi\mathcal{D}_x^{\alpha+1}y - \sum_{n=0}^{\infty}\binom{\alpha+1}{n}\mathcal{D}_x^{\alpha-n+1}y\mathcal{D}_x^n(\xi)\\
&= \mathcal{D}_x^\alpha\eta + \sum_{n=0}^{\infty}\binom{\alpha}{n}\mathcal{D}_x^{\alpha-n}y\mathcal{D}_x^{n+1}(\xi) + \xi\mathcal{D}_x^{\alpha+1}y - \sum_{n=-1}^{\infty}\binom{\alpha+1}{n+1}\mathcal{D}_x^{\alpha-n}y\mathcal{D}_x^{n+1}(\xi)\\
&= \mathcal{D}_x^\alpha\eta + \sum_{n=1}^{\infty}\binom{\alpha}{n}\mathcal{D}_x^{\alpha-n}y\mathcal{D}_x^{n+1}(\xi) + \xi\mathcal{D}_x^{\alpha+1}y + \binom{\alpha}{0}\mathcal{D}_x^\alpha y\mathcal{D}_x(\xi)\\
&\quad - \sum_{n=1}^{\infty}\binom{\alpha+1}{n+1}\mathcal{D}_x^{\alpha-n}y\mathcal{D}_x^{n+1}(\xi) - \binom{\alpha}{0}\xi\mathcal{D}_x^{\alpha+1}y - \binom{\alpha+1}{1}\mathcal{D}_x^\alpha y\mathcal{D}_x(\xi)\\
&= \mathcal{D}_x^\alpha\eta - \alpha\mathcal{D}_x^\alpha y\mathcal{D}_x(\xi) + \sum_{n=1}^{\infty}\left[\binom{\alpha}{n} - \binom{\alpha+1}{n+1}\right]\mathcal{D}_x^{\alpha-n}y\mathcal{D}_x^{n+1}(\xi).
\end{aligned} \tag{3.18}$$

Thus, the rest task is to determine the first term $\mathcal{D}_x^\alpha\eta$. In view of the generalized chain rule

$$\frac{d^m g(y(x))}{dx^m} = \sum_{k=0}^{m}\sum_{r=0}^{k}\binom{k}{r}\frac{1}{k!}[-y(x)]^r\frac{d^m}{dx^m}\left[(y(x))^{k-r}\right]\frac{d^k g(y)}{dy^k}, \tag{3.19}$$

and the generalized Leibniz rule with $f(x) = f(x, y(x))$ and $g(x) = 1$, we have

$$\begin{aligned}
\mathcal{D}_x^\alpha[f(x, y(x))] = \sum_{n=0}^{\infty}\sum_{m=0}^{n}\sum_{k=0}^{m}\sum_{r=0}^{k}\binom{\alpha}{n}\binom{n}{m}\binom{k}{r}\frac{1}{k!}\frac{x^{n-\alpha}}{\Gamma(n+1-\alpha)}\\
\times\left[-y(x)\right]^r\frac{d^m}{dx^m}\left[(y(x))^{k-r}\right]\frac{\partial^{n-m+k}f(x,y)}{\partial x^{n-m}\partial y^k}.
\end{aligned} \tag{3.20}$$

Some calculations on (3.20) and rearrangement of the resulting terms yields (see [154] for details of the proof)

$$\mathcal{D}_x^\alpha(f) = \partial_x^\alpha(f) + \partial_x^\alpha(yf_y) - y\partial_x^\alpha(f_y) + \mathfrak{g}, \tag{3.21}$$

where

$$\partial_x^\alpha(f) := \partial_x^\alpha[f(x,y)] = \frac{1}{\Gamma(1-\alpha)}\frac{\partial}{\partial x}\int_0^x \frac{f(t,y)}{(x-t)^\alpha}dt, \tag{3.22}$$

and

$$\begin{aligned}\mathfrak{g}(x,y) = &\sum_{n=0}^{\infty}\sum_{m=2}^{n}\sum_{k=2}^{m}\sum_{r=0}^{k-1}\binom{\alpha}{n}\binom{n}{m}\binom{k}{r}\frac{1}{k!}\frac{x^{n-\alpha}}{\Gamma(n+1-\alpha)}\\ &\times[-y(x)]^r\frac{d^m}{dx^m}\left[(y(x))^{k-r}\right]\frac{\partial^{n-m+k}f(x,y)}{\partial x^{n-m}\partial y^k}.\end{aligned} \tag{3.23}$$

As a consequence, for $k \geq 2$, we have

$$\mathfrak{g}|_{f_{yy}=0} = 0. \tag{3.24}$$

Similar to (3.21), one can write

$$\begin{aligned}&\mathcal{D}_x^\alpha(\eta) = \partial_x^\alpha(\eta) + \partial_x^\alpha(y\eta_y) - y\partial_x^\alpha(\eta_y) + \mu,\\ &\mathcal{D}_x^{n+1}(\xi) = \frac{\partial^{n+1}}{\partial x^{n+1}} + \sum_{m=1}^{n+1}\binom{n+1}{m}y^{(m)}\frac{\partial^{n-m+1}}{\partial x^{n-m+1}}\xi_y + \nu_n, n=1,2,...,\end{aligned} \tag{3.25}$$

where

$$\begin{aligned}\mu = &\sum_{n=0}^{\infty}\sum_{m=2}^{n}\sum_{k=2}^{m}\sum_{r=0}^{k-1}\binom{\alpha}{n}\binom{n}{m}\binom{k}{r}\frac{1}{k!}\frac{x^{n-\alpha}}{\Gamma(n+1-\alpha)}\\ &\times[-y(x)]^r\frac{d^m}{dx^m}\left[(y(x))^{k-r}\right]\frac{\partial^{n-m+k}\eta(x,y)}{\partial x^{n-m}\partial y^k},\end{aligned} \tag{3.26}$$

$$\begin{aligned}\nu_n = &\sum_{i=0}^{\infty}\sum_{j=2}^{i}\sum_{k=2}^{j}\sum_{r=0}^{k-1}\binom{n+1}{i}\binom{i}{j}\binom{k}{r}\frac{1}{k!}\frac{x^{i-(n+1)}}{\Gamma(i+1-(n+1))}\\ &\times[-y(x)]^r\frac{d^j}{dx^j}\left[(y(x))^{k-r}\right]\frac{\partial^{i-j+k}\xi(x,y)}{\partial x^{i-j}\partial y^k}, \quad n=1,2,\dots,\end{aligned} \tag{3.27}$$

with the results

$$\mu|_{\eta_{yy}=0} = 0, \quad \nu_n|_{\xi_{yy}=0} = 0, \quad n=1,2,.... \tag{3.28}$$

If we rewrite (3.18) by using the Eqs. (3.25), then we obtain an explicit form of η^α as:

$$
\begin{aligned}
\eta^\alpha &= \partial_x^\alpha(\eta) + \partial_x^\alpha(y\eta_y) - y\partial_x^\alpha(\eta_y) + \mu - \alpha\mathcal{D}_x^\alpha y(\xi_x + y'\xi_y) \\
&+ \left[\binom{\alpha}{1} - \binom{\alpha+1}{2}\right]\mathcal{D}_x^{\alpha-1} y \mathcal{D}_x^2\xi + \sum_{n=2}^{\infty}\left[\binom{\alpha}{n} - \binom{\alpha+1}{n+1}\right] \\
&\times \mathcal{D}_x^{\alpha-n} y \left[\frac{\partial^{n+1}}{\partial x^{n+1}}\xi + \sum_{m=1}^{n+1}\binom{n+1}{m} y^{(m)}\frac{\partial^{n-m+1}}{\partial x^{n-m+1}}\xi_y + \nu_n\right] \\
&= \partial_x^\alpha(\eta) + \sum_{n=0}^{\infty}\binom{\alpha}{n}\mathcal{D}_x^{\alpha-n} y \mathcal{D}_x^n \eta_y - y\partial_x^\alpha(\eta_y) + \mu - \alpha\xi_x\mathcal{D}_x^\alpha y - \alpha y'\xi_y\mathcal{D}_x^\alpha y \\
&+ \left[\alpha - \frac{\alpha(\alpha+1)}{2}\right]\mathcal{D}_x^{\alpha-1} y \mathcal{D}_x(\xi_x + y'\xi_y) + \sum_{n=2}^{\infty}\left[\binom{\alpha}{n} - \binom{\alpha+1}{n+1}\right] \\
&\times \mathcal{D}_x^{\alpha-n} y \left[\frac{\partial^{n+1}}{\partial x^{n+1}}\xi + \sum_{m=1}^{n+1}\binom{n+1}{m} y^{(m)}\frac{\partial^{n-m+1}}{\partial x^{n-m+1}}\xi_y + \nu_n\right] \\
&= \partial_x^\alpha(\eta) + \eta_y\mathcal{D}_x^\alpha y - y\partial_x^\alpha(\eta_y) + \alpha\mathcal{D}_x^{\alpha-1} y\mathcal{D}_x\eta_y + \mu + \sum_{n=2}^{\infty}\binom{\alpha}{n}\mathcal{D}_x^{\alpha-n} y \mathcal{D}_x^n\eta_y \\
&- \alpha\xi_x\mathcal{D}_x^\alpha y - \alpha y'\xi_y\mathcal{D}_x^\alpha y - \frac{\alpha(\alpha-1)}{2}\mathcal{D}_x^{\alpha-1} y\left[y'^2\xi_{yy} + 2y'\xi_{xy} + y''\xi_y + \xi_{xx}\right] \\
&+ \sum_{n=2}^{\infty}\left[\binom{\alpha}{n} - \binom{\alpha+1}{n+1}\right] \\
&\times \mathcal{D}_x^{\alpha-n} y \left[\frac{\partial^{n+1}}{\partial x^{n+1}}\xi + \sum_{m=1}^{n+1}\binom{n+1}{m} y^{(m)}\frac{\partial^{n-m+1}}{\partial x^{n-m+1}}\xi_y + \nu_n\right],
\end{aligned}
\tag{3.29}
$$

where

$$
\begin{aligned}
&\mathcal{D}_x\eta_y = \eta_{xy} + y'\eta_{yy}, \\
&\mathcal{D}_x^n\eta_y = \partial_x^n\eta_y + \sum_{m=1}^{n}\binom{n}{m} y^{(m)}\partial_x^{n-m}\eta_{yy} + \omega_n, \quad n = 2, 3, ...
\end{aligned}
\tag{3.30}
$$

and

$$
\begin{aligned}
\omega_n &= \sum_{i=0}^{\infty}\sum_{j=2}^{i}\sum_{k=2}^{j}\sum_{r=0}^{k-1}\binom{n}{i}\binom{i}{j}\binom{k}{r}\frac{1}{k!}\frac{x^{i-n}}{\Gamma(i+1-n)} \\
&\times [-y(x)]^r\frac{d^j}{dx^j}\left[(y(x))^{k-r}\right]\frac{\partial^{i-j+k}\eta_y(x,y)}{\partial x^{i-j}\partial y^k}, \quad n = 2, 3, ...
\end{aligned}
\tag{3.31}
$$

3.3 Symmetry group of FIDEs

In this section, we analyze the symmetry group of a FIDE of the form

$$\Delta := \mathcal{D}_x^\alpha y(x) - f(x, y(x)) - \mathcal{T}(y)(x) = 0, \tag{3.32}$$

where

$$\mathcal{T}(y)(x) = \int_0^x \mathcal{K}(x,t)y(t)dt, \tag{3.33}$$

x and y are independent and dependent variables, respectively, and the functions f and $\mathcal{K}$ are given smooth functions. According to the subjects of the previous section, the infinitesimal generator and its prolongation for the Eq.(3.32) can be written as

$$V = \xi\frac{\partial}{\partial x} + \eta\frac{\partial}{\partial y}, \tag{3.34}$$

and

$$Pr^{\mathcal{T},\alpha}V = \xi\frac{\partial}{\partial x} + \eta\frac{\partial}{\partial y} + \eta^\alpha\frac{\partial}{\partial \mathcal{D}_x^\alpha y} + P_{\mathcal{T}}(y)\frac{\partial}{\partial \mathcal{T}(y)}. \tag{3.35}$$

If we apply the prolongation (3.35) to (3.32) then we have

$$Pr^{\mathcal{T},\alpha}V(\Delta) = \eta^\alpha - P_{\mathcal{T}}(y) - f_x\xi - f_y\eta = 0, \tag{3.36}$$

where $P_{\mathcal{T}}(y)$ is defined by (3.12), i.e.,

$$P_{\mathcal{T}}(y)(x) = \int_0^x \left\{\left[\frac{\partial \mathcal{K}}{\partial x}\xi[x] + \frac{\partial \mathcal{K}}{\partial t}\xi[t]\right]y(t) + \mathcal{K}\eta[t] + \mathcal{K}y(t)D_t\xi[t]\right\}dt \tag{3.37}$$

with $\xi[x] := \xi(x, y(x))$ and $\eta[t] := \eta(t, y(t))$.
It is necessary to mention that the point transformations may change an integral equation to another integral equation. Thus, for some arbitrary function $c(x)$, the left side of Eq. (3.36) may be written as:

$$\eta^\alpha - P_{\mathcal{T}}(y) - f_x\xi - f_y\eta = c(x)\left(\mathcal{D}_x^\alpha y - f(x, y(x)) - \mathcal{T}(y)(x)\right). \tag{3.38}$$

Now, we assume that there are the functions $b(x)$ and $\varphi(x)$ such that

$$P_{\mathcal{T}}(y) = b(x)\mathcal{T}(y) + \varphi(x). \tag{3.39}$$

Then Eq. (3.38) can be rewritten as

$$\eta^\alpha - \varphi - f_x\xi - f_y\eta = c(x)(\mathcal{D}_x^\alpha y - f) + (b(x) - c(x))\mathcal{T}(y), \tag{3.40}$$

or equivalently, from (3.29):

$$\begin{aligned}
&\partial_x^\alpha(\eta) + \eta_y \mathcal{D}_x^\alpha y - y\partial_x^\alpha(\eta_y) + \alpha\mathcal{D}_x^{\alpha-1}y(\eta_{xy} + y'\eta_{yy}) + \mu \\
&+ \sum_{n=2}^{\infty}\binom{\alpha}{n}\mathcal{D}_x^{\alpha-n}y\left[\partial_x^n\eta_y + \sum_{m=1}^{n}\binom{n}{m}y^{(m)}\partial_x^{n-m}\eta_{yy} + \omega_n\right] \\
&- \alpha\xi_x\mathcal{D}_x^\alpha y - \alpha y'\xi_y\mathcal{D}_x^\alpha y - \frac{\alpha(\alpha-1)}{2}\mathcal{D}_x^{\alpha-1}y\left[y'^2\xi_{yy} + 2y'\xi_{xy} + y''\xi_y + \xi_{xx}\right] \\
&+ \sum_{n=2}^{\infty}\left[\binom{\alpha}{n} - \binom{\alpha+1}{n+1}\right]\mathcal{D}_x^{\alpha-n}y \\
&\times\left[\frac{\partial^{n+1}}{\partial x^{n+1}}\xi + \sum_{m=1}^{n+1}\binom{n+1}{m}y^{(m)}\frac{\partial^{n-m+1}}{\partial x^{n-m+1}}\xi_y + \nu_n\right] \\
&- \varphi - f_x\xi - f_y\eta = c(x)(\mathcal{D}_x^\alpha y - f) + (b(x) - c(x))\mathcal{T}(y). \qquad (3.41)
\end{aligned}$$

By equating the coefficients of $\mathcal{T}(y)$, $\mathcal{D}_x^\alpha y$, $\mathcal{D}_x^{\alpha-1}y$, $\mathcal{D}_x^{\alpha-n}y$, $y'\mathcal{D}_x^\alpha y$, $y'\mathcal{D}_x^{\alpha-1}y$, $y''\mathcal{D}_x^{\alpha-1}y$, $y'^2\mathcal{D}_x^{\alpha-1}y$, ..., with zero in (3.41), we obtain the following overdetermined system of equations which are called determining equations:

(A1) $\xi_y = 0$,

(A2) $\eta_{yy} = 0$,

(A3) $\eta_y - \alpha\xi_x = c(x)$,

(A4) $\alpha\eta_{xy} - \frac{\alpha(\alpha-1)}{2}\xi_{xx} = 0$,

(A5) $\binom{\alpha}{n}\partial_x^n\eta_y + \left[\binom{\alpha}{n} - \binom{\alpha+1}{n+1}\right]\frac{\partial^{n+1}}{\partial x^{n+1}}\xi = 0, \qquad n = 2, 3, ...,$

(A6) $b(x) - c(x) = 0$,

(A7) $\partial_x^\alpha\eta - y\partial_x^\alpha\eta_y - f_x\xi - f_y\eta - \varphi + cf = 0$.

Eq.(A1) implies that $\xi = \xi(x)$. So according to (3.28) all terms of ν_n vanish. Also, Eq. (A2) implies that all terms of μ and ω_n in (3.26) and (3.31) must be vanished. Solving $\eta_{yy} = 0$ yields

$$\eta(x, y) = e(x)y + h(x), \qquad (3.42)$$

where $e(x)$ and $h(x)$ are unknown functions to be determined.
From (A4) and (3.42), we have

$$\eta_y = e(x) = \frac{\alpha - 1}{2}\xi_x + q, \qquad (3.43)$$

where q is a constant. Substituting this into (A5) yields

$$\begin{aligned}
&\binom{\alpha}{n}\partial_x^n e(x) + \left[\binom{\alpha}{n} - \binom{\alpha+1}{n+1}\right]\frac{\partial^{n+1}}{\partial x^{n+1}}\xi \\
&= \binom{\alpha}{n}\partial_x^n\left[\frac{\alpha-1}{2}\xi_x + q\right] + \left[\binom{\alpha}{n} - \binom{\alpha+1}{n+1}\right]\frac{\partial^{n+1}}{\partial x^{n+1}}\xi \\
&= \binom{\alpha}{n}\partial_x^n\left[\frac{\alpha-1}{2}\xi_x\right] + \left[\binom{\alpha}{n} - \binom{\alpha+1}{n+1}\right]\frac{\partial^{n+1}}{\partial x^{n+1}}\xi \\
&= \frac{\alpha-1}{2}\binom{\alpha}{n}\frac{\partial^{n+1}}{\partial x^{n+1}}\xi + \left[\binom{\alpha}{n} - \binom{\alpha+1}{n+1}\right]\frac{\partial^{n+1}}{\partial x^{n+1}}\xi \\
&= \left[\frac{\alpha+1}{2}\binom{\alpha}{n} - \binom{\alpha+1}{n+1}\right]\frac{\partial^{n+1}}{\partial x^{n+1}}\xi = 0, \qquad n = 2, 3, ..., \qquad (3.44)
\end{aligned}$$

and so for $n = 2$, we have $\xi'''(x) = 0$, which has the general solution

$$\xi(x) = c_1x^2 + c_2x + c_3, \qquad (3.45)$$

where c_1, c_2 and c_3 are arbitrary constants. By substituting (3.45) into (3.43), we find

$$e(x) = c_1(\alpha - 1)x + c_4, \qquad c_4 = \frac{\alpha-1}{2}c_2 + q. \qquad (3.46)$$

Now, substituting $\eta(x, y) = e(x)y + h(x)$ in (A7) implies

$$\xi f_x + \eta f_y = cf + \partial_x^\alpha h - \varphi. \qquad (3.47)$$

On the other hand, from (3.37), (3.42) and (3.39) it can be written:

$$\begin{aligned}
P_{\mathcal{T}}(y) &= \int_0^x \left\{\left[\frac{\partial\mathcal{K}(x,t)}{\partial x}\xi[x] + \frac{\partial\mathcal{K}(x,t)}{\partial t}\xi[t]\right]y(t) + \mathcal{K}(x,t)\eta[t] + \mathcal{K}(x,t)y(t)\xi'(t)\right\}dt \\
&= \int_0^x \left[\frac{\partial\mathcal{K}(x,t)}{\partial x}\xi[x] + \frac{\partial\mathcal{K}(x,t)}{\partial t}\xi[t] + \mathcal{K}(x,t)e(t) + \mathcal{K}(x,t)\xi'(t)\right]y(t)dt \\
&+ \int_0^x \mathcal{K}(x,t)h(t)dt. \qquad (3.48)
\end{aligned}$$

Then, due to (3.39), we set $\varphi(x) = \int_0^x \mathcal{K}(x,t)h(t)dt$ and determine the functions $\xi(x)$ and $e(x)$ in such a way that

$$\frac{\partial\mathcal{K}(x,t)}{\partial x}\xi[x] + \frac{\partial\mathcal{K}(x,t)}{\partial t}\xi[t] + \mathcal{K}(x,t)e(t) + \mathcal{K}(x,t)\xi'(t) = \mathcal{K}(x,t)b(x). \quad (3.49)$$

By using (3.48), Eq.(3.47) can be rewritten as follows:

$$\xi f_x + \eta f_y = cf + \partial_x^\alpha h - \int_0^x \mathcal{K}(x,t)h(t)dt. \qquad (3.50)$$

From Eq. ($A6$), we have $b(x) = c(x)$. Substituting η_y and ξ_x from (3.43) and (3.45) in Eq. ($A3$), we get

$$b(x) = c(x) = -c_1(\alpha + 1)x - \frac{\alpha + 1}{2}c_2 + q. \tag{3.51}$$

Eventually, we achieve the general point transformation group which are infinitesimals of the infinitesimal generator V(3.34), determined by the following relations:

$$\begin{aligned} \xi &= c_1x^2 + c_2x + c_3, \\ \eta &= \Big(c_1(\alpha - 1)x + \frac{\alpha - 1}{2}c_2 + q\Big)y + h(x), \end{aligned} \tag{3.52}$$

where c_1, c_2, c_3 and q are arbitrary constants and $h(x)$ is an arbitrary function. In addition to the results (3.52), we obtain two main differential equations depending on the symmetries analysis for the equation (3.32). The first one is a partial differential equation obtained from (3.49) in terms of the kernel function $\mathcal{K}(x, t)$:

$$(c_1x^2+c_2x+c_3)\frac{\partial \mathcal{K}}{\partial x}+(c_1t^2+c_2t+c_3)\frac{\partial \mathcal{K}}{\partial t}+(\alpha+1)(c_1(t+x)+c_2)\mathcal{K} = 0, \tag{3.53}$$

and the second one is the fractional integro-differential equation in terms of $f(x, y(x))$:

$$\begin{aligned} &(c_1x^2 + c_2x + c_3)f_x + \left[\Big(c_1(\alpha - 1)x + \frac{\alpha - 1}{2}c_2 + q\Big)y + h(x)\right] f_y \\ &+ \left[c_1(\alpha + 1)x + \frac{\alpha + 1}{2}c_2 - q\right] f = \partial_x^\alpha h - \int_0^x \mathcal{K}(x, t)h(t)dt. \end{aligned} \tag{3.54}$$

3.4 Kernel function, free term and related symmetry group of the FIDEs

In this section, we present the symmetries of the FIDE (3.32) based on the kernel function and free term.

3.4.1 General conditions for $\mathcal{K}$ and f

The general conditions for the kernel function and free term, according to the Eqs. (3.53) and (3.54), are given by:

$$c_1 = c_2 = c_3 = q = 0, \tag{3.55}$$

and the generator is:

$$V = h(x)\frac{\partial}{\partial y}, \tag{3.56}$$

where $h(x)$ is the solution of the following equation:

$$h(x)f_y(x,y) = \partial_x^\alpha h(x) - \int_0^x \mathcal{K}(x,t)h(t)dt. \tag{3.57}$$

3.4.2 Some special cases

1. If we set

$$c_1 = c_2 = q = 0, \qquad h = c_3 \neq 0, \tag{3.58}$$

then from Eq. (3.53), we have

$$\frac{\partial \mathcal{K}}{\partial x} + \frac{\partial \mathcal{K}}{\partial t} = 0, \tag{3.59}$$

and Eq. (3.54) yields:

$$\frac{\partial f}{\partial x} + \frac{\partial f}{\partial y} = l(x), \tag{3.60}$$

where

$$l(x) = \frac{1}{\Gamma(1-\alpha)x^\alpha} - \int_0^x \mathcal{K}(x,t)dt. \tag{3.61}$$

By solving the above equations, we get

$$\mathcal{K}(x,t) = \mathcal{F}_1(t-x), \tag{3.62}$$

$$f(x,y) = \mathcal{F}_2(y-x) + \int l(x)dx. \tag{3.63}$$

As an example, for $\mathcal{K}(x,t) = t - x$ and $\alpha = 1/2$ we have

$$f(x,y) = \mathcal{F}_2(y-x) + \frac{2}{\sqrt{\pi}}\sqrt{x} + \frac{x^3}{6}. \tag{3.64}$$

In this case, we find

$$\xi(x) = c_3, \qquad \eta(x,y) = c_3 \tag{3.65}$$

and the corresponding generator

$$V = \frac{\partial}{\partial x} + \frac{\partial}{\partial y}. \tag{3.66}$$

2. Set $c_1 = c_3 = h = q = 0$ and $c_2 \neq 0$. Then for $\alpha = \frac{1}{2}$, we obtain from (3.53) and (3.54) the following equation:

$$x\frac{\partial \mathcal{K}}{\partial x} + t\frac{\partial \mathcal{K}}{\partial t} + \frac{3}{2}\mathcal{K} = 0, \tag{3.67}$$

$$x\frac{\partial f}{\partial x} - \frac{1}{4}y\frac{\partial f}{\partial y} + \frac{3}{4}f = 0. \tag{3.68}$$

Solutions of Eqs. (3.67)-(3.68) are in the form:

$$\mathcal{K}(x,t) = \frac{\mathcal{F}_1(\frac{t}{x})}{x^{3/2}}, \quad f(x,y) = \frac{\mathcal{F}_2(yx^{1/4})}{x^{3/4}}, \tag{3.69}$$

with infinitesimals:

$$\xi(x) = c_2 x, \qquad \eta(x,y) = -\frac{1}{4}c_2 y \tag{3.70}$$

and the corresponding generator:

$$V = x\frac{\partial}{\partial x} - \frac{1}{4}y\frac{\partial}{\partial y}. \tag{3.71}$$

3. Similar to the previous cases, for $c_1 = c_2 = 0$ and $c_3 = q = h \neq 0$, from (3.53) we obtain

$$\frac{\partial \mathcal{K}}{\partial x} + \frac{\partial \mathcal{K}}{\partial t} = 0. \tag{3.72}$$

Hence, the kernel function can be written as:

$$\mathcal{K}(x,t) = \mathcal{F}_1(t-x) \tag{3.73}$$

and Eq. (3.54) concludes:

$$\frac{\partial f}{\partial x} + (y+1)\frac{\partial f}{\partial y} - f = l(x), \tag{3.74}$$

with

$$l(x) = \frac{1}{\Gamma(1-\alpha)x^{\alpha}} - \int_0^x \mathcal{K}(x,t)dt, \tag{3.75}$$

which has the solution

$$f(x,y) = \left(\int l(x)\exp(-x)dx + \mathcal{F}_2((y+1)\exp(-x))\right)\exp(x). \tag{3.76}$$

As an example, for $\mathcal{K}(x,t) = t - x$ and $\alpha = 1/2$, we have

$$f(x,y) = -\frac{x^2}{2} - x - 1 + \frac{\exp(x)\sqrt{\pi}}{\Gamma(1/2)}erf(\sqrt{x}) + \exp(x)F_2((y+1)\exp(-x)), \tag{3.77}$$

where $erf(\sqrt{x})$ is the error function defined by

$$erf(x) = \frac{2\int_0^x \exp(-t^2)dt}{\sqrt{\pi}}. \tag{3.78}$$

In this case, the infinitesimal functions take the form

$$\xi(x) = c_3, \qquad \eta(x,y) = c_3(y+1) \tag{3.79}$$

with corresponding generator:

$$V = \frac{\partial}{\partial x} + (y+1)\frac{\partial}{\partial y}. \tag{3.80}$$

4. For $c_1 = c_3 = q = 0$, $c_2 = h \neq 0$ and $\alpha = \frac{1}{2}$, from (3.53) we obtain

$$x^2\frac{\partial \mathcal{K}}{\partial x} + t^2\frac{\partial \mathcal{K}}{\partial t} + \frac{3}{2}(t+x)\mathcal{K} = 0, \tag{3.81}$$

with the solution

$$\mathcal{K}(x,t) = \frac{\mathcal{F}_1(-\frac{t-x}{tx})(-\frac{t-x}{t}+1)^{\frac{3}{2}}}{x^3}. \tag{3.82}$$

From (3.54), we have

$$x\frac{\partial f}{\partial x} + \left[-\frac{1}{4}y + 1\right]\frac{\partial f}{\partial y} + \frac{3}{4}f = l(x), \tag{3.83}$$

where

$$l(x) = \frac{1}{\Gamma(1/2)\sqrt{x}} - \int_0^x \mathcal{K}(x,t)dt, \tag{3.84}$$

and solution

$$f(x,y) = \frac{\int \frac{l(x)}{x^{1/4}}dx + \mathcal{F}_2\left(yx^{1/4} - 4x^{1/4}\right)}{x^{3/4}}. \tag{3.85}$$

As an example, for $\mathcal{K}(x,t) = \frac{(-\frac{t-x}{tx})(-\frac{t-x}{t}+1)^{\frac{3}{2}}}{x^3}$ and

$$f(x,y) = \frac{-2+\frac{4}{\sqrt{\pi}}}{\sqrt{x}} + \frac{\mathcal{F}_2(yx^{1/4} - 4x^{1/4})}{x^{3/4}}, \tag{3.86}$$

the infinitesimal functions take the form

$$\xi(x) = c_2 x, \qquad \eta(x,y) = -\frac{1}{4}c_2 y + c_2, \tag{3.87}$$

with the corresponding generator

$$V = x\frac{\partial}{\partial x} + (-\frac{1}{4}y + 1)\frac{\partial}{\partial y}. \tag{3.88}$$

5. For the continuous kernel $\mathcal{K} := \mathcal{K}(x,t)$, we consider

$$c_1 = c_2 = c_3 = 0, \; h = q \neq 0. \tag{3.89}$$

Substituting in Eq. (3.54) yields the differential equation

$$(y+1)f_y - f = \frac{1}{\Gamma(1-\alpha)\sqrt{x}} - \rho(x), \tag{3.90}$$

where $\rho(x) = \int_0^x \mathcal{K}(x,t)dt$. The exact solution of Eq. (3.90) has the following form:

$$f(x,y) = F(x)y + F(x) + \rho(x) - \frac{1}{\Gamma(1-\alpha)\sqrt{x}}, \tag{3.91}$$

where $F(x)$ is an arbitrary function.
In this case, the infinitesimal functions are

$$\xi(x) = 0, \qquad \eta(x,y) = q(y+1), \tag{3.92}$$

and the corresponding generator:

$$V = (y+1)\frac{\partial}{\partial y}. \tag{3.93}$$

Now, we consider some examples of FIDEs with corresponding invariant solutions.

Example 1: Consider a FIDE as follows:

$$\mathcal{D}_x^{1/2}y(x) = -\frac{x^2}{2} - x + \exp(x)erf(\sqrt{x}) + y(x) + \int_0^x (t-x)y(t)dt. \tag{3.94}$$

From the previous section, corresponding infinitesimal generator is:

$$V = \frac{\partial}{\partial x} + (y+1)\frac{\partial}{\partial y}. \tag{3.95}$$

From the characteristics equation $\frac{dx}{1} = \frac{dy}{y+1}$, we obtain the invariant

$$y = \exp(x) - 1, \tag{3.96}$$

which satisfies the Eq. (3.94).

Example 2: Let us to consider another FIDE of the form

$$\mathcal{D}_x^{1/2}y(x) = \frac{-2 + \frac{4}{\sqrt{\pi}}}{\sqrt{x}} + \frac{7\Gamma(3/4)}{11\Gamma(1/4)}\frac{y(x)}{\sqrt{x}} + \int_0^x \frac{t}{x^{5/2}}y(t)dt. \tag{3.97}$$

From previous sections, it can be obtained:

$$V = x\frac{\partial}{\partial x} + \left(1 - \frac{1}{4}y\right)\frac{\partial}{\partial y}. \tag{3.98}$$

The characteristic equation $\frac{dx}{x} = \frac{dy}{-(1/4)y+1}$ concludes the solution of Eq. (3.97) as

$$y(x) = 4(1 - x^{-1/4}). \tag{3.99}$$

Example 3: Assuming

$$\mathcal{D}_x^{1/2} y(x) = 2x - x\cos(x) - \frac{1}{\sqrt{\pi}\sqrt{x}} + yx + \int_0^x x\sin(t)y(t)dt, \tag{3.100}$$

and according to the fifth case, its symmetry group is

$$V = (y+1)\frac{\partial}{\partial y}. \tag{3.101}$$

The characteristic equation $\frac{dx}{0} = \frac{dy}{y+1}$ gives

$$y = -1, \tag{3.102}$$

which is also a solution of Eq. (3.100).

Chapter 4

Nonclassical Lie symmetry analysis to fractional differential equations

This chapter deals with the improvement of nonclassical Lie symmetries to fractional differential equations. In previous chapters the classical Lie symmetries were considered for fractional differential and integro-differential equations. But, here, we discuss the vector fields which are not accessible from classical ones.

Let us introduce the notation, concepts and lemmas which are important in this chapter.

Proposition 4.0.1 [139] *(Fractional Taylor's expansions) Let f be analytical in $(-h, h)$ for some $h > 0$, and let $\alpha > 0$, $\alpha \notin \mathbb{N}$. Then*

$$I^{\alpha} f(t) = \sum_{n=0}^{\infty} \frac{(-1)^n t^{n+\alpha}}{n!(\alpha+n)\Gamma(\alpha)} f^{(n)}(t),$$

and

$$D_t^{\alpha} f(t) = \sum_{n=0}^{\infty} \binom{\alpha}{n} \frac{t^{n-\alpha}}{\Gamma(n+1-\alpha)} f^{(n)}(t), \tag{4.1}$$

for $0 < t < h/2$ where $f^{(n)}(t)$ denotes the n-th derivative of f at t and $D_t^{\alpha} := {}^{RL}D_t^{\alpha}$ is the Riemann–Liouville fractional derivative of order $\alpha > 0$.

Proposition 4.0.2 [139] *(Leibniz's formula for Riemann–Liouville fractional operator) Let $\alpha > 0$ and assume that f and g are analytical on $(-h, h)$ with some $h > 0$. Then,*

$$D_t^{\alpha}[fg](t) = \sum_{n=0}^{\infty} \binom{\alpha}{n} f^{(n)}(t) D_t^{\alpha-n} g(t),$$

for $0 < t < h/2$.

Remark 4.0.1 *In the case that the function f is multi-variable, $f(x, t, u)$, $D_t^{\alpha} f(x, t, u)$ stands for the partial fractional derivative of f w.r.t. t where the other variables, x and u, are constant. In this stage, if $u = u(x, t)$, $\mathcal{D}_t^{\alpha} f$ and $\mathcal{D}_t f$ denote the total fractional derivative and the total derivative of f, w.r.t. t, respectively.*

Now we give a brief review of the Lie symmetry analysis concerning the fractional differential operator which is a nonlocal operator. Let

$$\begin{aligned}\bar{x} &= x + \varepsilon\xi(x,t,u) + O(\varepsilon^2),\\ \bar{t} &= t + \varepsilon\tau(x,t,u) + O(\varepsilon^2),\\ \bar{u} &= u + \varepsilon\varphi(x,t,u) + O(\varepsilon^2),\end{aligned} \tag{4.2}$$

be a one-parameter group of transformations G on an open subset $M \subset X \times U \simeq R^3$ with coordinate (x,t,u) where ε is the group parameter. Then its infinitesimal generator has the following form

$$V = \xi(x,t,u)\frac{\partial}{\partial x} + \tau(x,t,u)\frac{\partial}{\partial t} + \varphi(x,t,u)\frac{\partial}{\partial u}. \tag{4.3}$$

To obtain invariance of a differential equation involving a fractional derivative operator, we have to give the prolongation of these operators under the group of transformations. Now we have some results of [66, 139].

Theorem 21 *Let $u(\cdot,\cdot)$ be an analytical function, and let $\alpha > 0$, $\alpha \notin \mathbb{N}$, $m-1 < \alpha \leq m$. Then*

$$D_{\bar{t}}^{\alpha}\bar{u}(\bar{x},\bar{t}) = D_t^{\alpha}u(x,t) + \varepsilon\varphi^{(\alpha,t)}(x,t,u) + O(\varepsilon^2),$$

where

$$\begin{aligned}\varphi^{(\alpha,t)} =& \mathcal{D}_t^{\alpha}\varphi + \xi D_t^{\alpha}u_x - \mathcal{D}_t^{\alpha}(\xi u_x) - \mathcal{D}_t^{\alpha+1}(\tau u)\\ &+ \mathcal{D}_t^{\alpha}(\mathcal{D}_t(\tau)u) + \tau D_t^{\alpha+1}u.\end{aligned}$$

To keep the lower limit of integral in the Riemann-Liouville operator, invariant under the group of transformations (4.2), we have to suppose

$$\tau(x,t,u(x,t))|_{t=0} = 0.$$

Theorem 22 *Let $\alpha > 0$ and suppose*

$$F(x,t,u,u_x,u_{xx},\ldots,D_t^{\alpha}u) = 0 \tag{4.4}$$

is a time-fractional differential equation defined over $M \subset X \times U \simeq R^3$. Let G be a local group of transformations acting on M as in (4.2), and

$$Pr^{(\alpha,t)}V(F) = 0,$$

whenever $F(x,t,u,u_x,u_{xx},\ldots,D_t^{\alpha}u) = 0$, for every infinitesimal generator V of G, where

$$Pr^{(\alpha,t)}V = V + \varphi^x\frac{\partial}{\partial u_x} + \varphi^{xx}\frac{\partial}{\partial u_{xx}} + \cdots + \varphi^{(\alpha,t)}\frac{\partial}{\partial D_t^{\alpha}u},$$

with $\varphi^{(\alpha,t)}$ as in Theorem 21 and

$$\begin{aligned}
\varphi^{x} &= \mathcal{D}_x\varphi - \mathcal{D}_x(\xi)u_x - \mathcal{D}_x(\tau)u_t,\\
\varphi^{xx} &= \mathcal{D}_x(\varphi^{x}) - \mathcal{D}_x(\xi)u_{xx} - \mathcal{D}_x(\tau)u_{xt},\\
\varphi^{xxx} &= \mathcal{D}_x(\varphi^{xx}) - \mathcal{D}_x(\xi)u_{xxx} - \mathcal{D}_x(\tau)u_{xxt},\\
&\dots
\end{aligned}$$

Then G is a symmetry group of (4.4).

Remark 4.0.2 *Using the Leibniz's formula for Riemann–Liouville fractional operator, we can write*

$$\begin{aligned}
\varphi^{(\alpha,t)} =& \mathcal{D}_t^{\alpha}\varphi - \sum_{n=1}^{\infty}\left(\begin{array}{c}\alpha\\ n\end{array}\right)\mathcal{D}_t^{n}\xi D_t^{\alpha-n}u_x\\
&- \sum_{n=1}^{\infty}\left(\begin{array}{c}\alpha\\ n\end{array}\right)\mathcal{D}_t^{n}\tau D_t^{\alpha+1-n}u,
\end{aligned} \tag{4.5}$$

where for $\alpha > 0$, $\mathcal{D}_t^{\alpha}\varphi$ can be computed using (4.1) and the chain rule,

$$\begin{aligned}
\mathcal{D}_t^{\alpha}\varphi(x,t,u(x,t)) &= \sum_{n=0}^{\infty}\binom{\alpha}{n}\frac{t^{n-\alpha}}{\Gamma(n+1-\alpha)}\\
\times\mathcal{D}_t^{n}\varphi(x,t,u(x,t)) &= D_t^{\alpha}\varphi + \varphi_u D_t^{\alpha}u - uD_t^{\alpha}\varphi_u\\
&+\sum_{n=1}^{\infty}\binom{\alpha}{n}D_t^{n}\varphi_u D_t^{\alpha-n}u + \mu,
\end{aligned}$$

with

$$\begin{aligned}
\mu =& \sum_{n=2}^{\infty}\left(\begin{array}{c}\alpha\\ n\end{array}\right)\frac{t^{n-\alpha}}{\Gamma(n+1-\alpha)}\sum_{m=2}^{n}\sum_{k=2}^{m}\sum_{r=0}^{k-1}\left(\begin{array}{c}n\\ m\end{array}\right)\left(\begin{array}{c}k\\ r\end{array}\right)\\
&\times\frac{1}{k!}(-u)^{r}D_t^{m}u^{k-r}D_t^{n-m}\left(D_u^{k}\varphi\right),
\end{aligned} \tag{4.6}$$

where $\varphi_u = \frac{\partial\varphi}{\partial u}$ and $D_t^{m} = \frac{\partial^{m}}{\partial t^{m}}$.

4.1 General solutions extracted from invariant surfaces to fractional differential equations

Theorem 22, when it is coupled with formula (4.5), provides an effective computational procedure for obtaining invariance of a fractional differential equation of the form (4.4). Due to the prolongation formula of the fractional derivative operators, the resulting determining equation is too complicated to be solved for infinitesimal and the classical method is obtained by setting the coefficients of u_x, u_x^2, etc., to zero. In this way, we lose some invariance. The nonclassical method of reduction was devised originally by Bluman and Cole in 1969, to find new exact solutions of the heat equation. Here we provide a brief conceptual background of nonclassical methods to fractional differential equation to obtain more symmetries. The main idea behind this method is to insert a surface invariant condition such that by combining this equation with the original determining equation, we deduce a nonlinear determining equation for infinitesimals. A new determining equation can be solved for infinitesimals too.

Let (4.4) be a time-fractional differential equation defined over $M \subset X \times U \simeq R^3$. Let G be a local group of transformations acting on M as in (4.2). To derive the solutions $u = u(x,t)$ of Eq. (4.4) such that they are invariant under group transformation (4.2), we have to set

$$\Lambda : \xi u_x + \tau u_t - \varphi = 0.$$

For the nonclassical method, we seek invariance of both the original equations together with the invariant surface condition. Then we have

$$Pr^{(\alpha)}V(F)|_{F=0,\Lambda=0} = 0, \tag{4.7}$$

$$Pr^{(\alpha)}V(\Lambda)|_{F=0,\Lambda=0} = 0. \tag{4.8}$$

In this case for every ξ, τ and φ the Eq. (4.8) is identically satisfied when $\Lambda = 0$. Indeed, we only consider the condition

$$\Lambda : \xi(x,t,u)u_x + \tau(x,t,u)u_t - \varphi(x,t,u) = 0, \tag{4.9}$$

along with Eq. (4.7). Without loss of generality, in practice we consider the cases $\xi = 1$ and $\xi = 0$. We notice that for time-fractional differential equations the infinitesimals with $\tau(x,t,u(x,t))|_{t=0} \neq 0$ are omitted.

At the end of this section we give an idea for construction of the solutions to an equation using a new invariant condition. For a given invariance to a fractional partial differential equation, reduction to a solvable fractional ordinary differential equation is not straightforward to capture invariant solutions. We are now ready to state the idea of finding the structure of solutions to fractional differential equations for which the invariance has been found. Let Eq. (4.4) admit the group of transformation (4.2) with infinitesimal generator (4.3). If the solution $g(x,t,u) = u(x,t) - u = 0$ is invariant, then it satisfies the surface condition (4.9). However, if there exists a class of solutions $g(x,t,u) = c$ to Eq. (4.4) admitting group (4.2), and $g(\bar{x},\bar{t},\bar{u}) = \bar{c}$ defines a solution too, then we have

$$Vg(x,t,u) = 1. \tag{4.10}$$

We merge this equation and the original equation to obtain the invariant class of solutions. For instance we have the following example.

Example 1 *Consider the fractional diffusion equation*

$$D_t^\alpha u + bu_{xx} = 0, \quad 0 < \alpha \leq 1, \tag{4.11}$$

where b is constant, and let $V = t^{\alpha-1}\frac{\partial}{\partial u}$ be the infinitesimal generator of (4.11). Then, according to (4.10), we can write

$$t^{\alpha-1} g_u = 1.$$

The solution of this equation can be written in the form

$$g(x,t,u) = t^{1-\alpha}u + t^{1-\alpha}w(x,t) = c_1,$$

so

$$u = c_1 t^{\alpha-1} - w(x,t).$$

For example if $w(x,t) = h(x)f(t)$, therefore

$$u = c_1 t^{\alpha-1} - h(x)f(t).$$

Substituting this solution into Eq. (4.11) and letting $h''(x) + kh(x) = 0$ where k is arbitrary constant, the reduced FODE is

$$D_t^\alpha f(t) - bkf(t) = 0.$$

The solution of this equation can be written in terms of Mittag–Leffler functions

$$f(t) = c_2 t^{\alpha-1} E_{\alpha,\alpha}(bkt^\alpha),$$

thus

$$u(x,t) = c_1 t^{\alpha-1} - h(x) t^{\alpha-1} E_{\alpha,\alpha}(bkt^\alpha)$$

is a family of solutions (4.11). As an example

$$u(x,t) = c_1 t^{\alpha-1} + (c_2 \cos x + c_3 \sin x) t^{\alpha-1} E_{\alpha,\alpha}(bkt^\alpha)$$

is a family of solutions (4.11).

Now, let us study the following nonlinear equation

$$\Delta : D_t^\alpha u + auu_x + bu_{xx} + cu_{xxx} + du(1-u) = 0, \tag{4.12}$$

where a, b, c, d are constants and $0 < \alpha \leq 1$. Let V (4.3) be an infinitesimal generator of the group G. It is prolonged as follows:

$$\begin{aligned} Pr^{(\alpha,t)}V =& V + \varphi^x \frac{\partial}{\partial u_x} + \varphi^{xx}\frac{\partial}{\partial u_{xx}} + \varphi^{xxx}\frac{\partial}{\partial u_{xxx}} \\ &+ \varphi^{(\alpha,t)}\frac{\partial}{\partial D_t^\alpha u}. \end{aligned}$$

A one-parameter Lie group of transformations is admitted by Eq. (4.12) if and only if

$$Pr^{(\alpha,t)}V(\Delta)|_{\Delta=0} = 0.$$

We find that ξ, τ and φ must satisfy the symmetry conditions

$$\varphi^{(\alpha,t)} + a\varphi u_x + a\varphi^x u + b\varphi^{xx} + c\varphi^{xxx} + d\varphi(1-2u) = 0,$$

whenever u satisfies (4.12). After substituting $\varphi^x, \varphi^{xx}, \varphi^{xxx}, \varphi^{(\alpha,t)}$ and replacing $D_t^\alpha u$ by $-auu_x - bu_{xx} - cu_{xxx} - du(1-u)$, we have the following

$$\begin{aligned}
&a\varphi_x u + b\varphi_{xx} + c\varphi_{xxx} + d[\varphi + (\alpha\tau_t - 2\varphi - \varphi_u)u \\
&+ (\varphi_u - \alpha\tau_t)u^2] + [a(\varphi + (\alpha\tau_t - \xi_x)u) + b(2\varphi_{xu} - \xi_{xx}) \\
&+ c(3\varphi_{xxu} - \xi_{xxx})]u_x + [-a\xi_u u + b(\varphi_{uu} - 2\xi_{xu}) \\
&+ 3c(\varphi_{xuu} - \xi_{xxu})]u_x^2 + [-b\xi_{uu} + c(\varphi_{uuu} - 3\xi_{xuu})]u_x^3 \\
&- c\xi_{xxx}u_x^4 - [a\tau_x u + b\tau_{xx} + c\tau_{xxx} - d\alpha\tau_u u(1-u)]u_t \\
&+ [a(\alpha-1)\tau_u u - 2b\tau_{xu} - 3c\tau_{xxu}]u_x u_t - c\tau_{uuu}u_x^3 u_t \\
&- [b\tau_{uu} + 3c\tau_{xuu}]u_x^2 u_t + [-3b\xi_u + 3c(\varphi_{uu} - 3\xi_{xu})]u_x u_{xx} \\
&+ [b(\alpha\tau_t - 2\xi_x) + 3c(\varphi_{xu} - \xi_{xx})]u_{xx} - 6c\xi_{uu}u_x^2 u_{xx} \\
&+ [b(\alpha-1)\tau_u - 3c\tau_{xu}]u_{xx}u_t - [2b\tau_x + 3c\tau_{xx}]u_{xt} \\
&- [2b\tau_u + 6c\tau_{xu}]u_{xt}u_x + c(\alpha\tau_t - 3\xi_x)u_{xxx} - 3c\tau_x u_{xxt} \\
&- 3c\tau_{uu}u_x^2 u_{xt} - 3c\xi_u u_{xx}^2 - 3c\tau_u u_{xxt}u_x - 3c\tau_u u_{xt}u_{xx} \\
&- 3c\tau_{uu}u_{xx}u_x u_t - 4c\xi_u u_{xxx}u_x + c(\alpha-1)\tau_u u_{xxx}u_t \\
&+ D_t^\alpha \varphi - uD_t^\alpha \varphi_u - \sum_{n=1}^{\infty}\binom{\alpha}{n}\mathcal{D}_t^n \xi D_t^{\alpha-n}u_x + \mu \\
&- \sum_{n=1}^{\infty}[\binom{\alpha}{n+1}\mathcal{D}_t^{n+1}\tau - \binom{\alpha}{n}D_t^n \varphi_u]D_t^{\alpha-n}u = 0,
\end{aligned} \tag{4.13}$$

where μ is defined as (4.6). We obtain the determining equations for the classical symmetry group as follows:

$$\begin{aligned}
&a\varphi_x u + b\varphi_{xx} + c\varphi_{xxx} + d[\varphi + (\alpha\tau_t - 2\varphi - \varphi_u)u \\
&+ (\varphi_u - \alpha\tau_t)u^2] = 0, \\
&a(\varphi + (\alpha\tau_t - \xi_x)u) + b(2\varphi_{xu} - \xi_{xx}) \\
&c(3\varphi_{xxu} - \xi_{xxx}) = 0, \\
&- a\xi_u u + b(\varphi_{uu} - 2\xi_{xu}) + 3c(\varphi_{xuu} - \xi_{xxu}) = 0, \\
&- b\xi_{uu} + c(\varphi_{uuu} - 3\xi_{xuu}) = 0, \qquad c\xi_{xxx} = 0, \\
&a\tau_x u + b\tau_{xx} + c\tau_{xxx} - d\alpha\tau_u u(1-u) = 0, \\
&a(\alpha-1)\tau_u u - 2b\tau_{xu} - 3c\tau_{xxu} = 0, \qquad c\tau_{uuu} = 0, \\
&b\tau_{uu} + 3c\tau_{xuu} = 0, \quad -3b\xi_u + 3c(\varphi_{uu} - 3\xi_{xu}) = 0, \\
&b(\alpha\tau_t - 2\xi_x) + 3c(\varphi_{xu} - \xi_{xx}) = 0, \qquad c\xi_{uu} = 0, \\
&b(\alpha-1)\tau_u - 3c\tau_{xu} = 0, \qquad 2b\tau_x + 3c\tau_{xx} = 0, \\
&2b\tau_u + 6c\tau_{xu} = 0, \qquad c(\alpha\tau_t - 3\xi_x) = 0, \\
&c\tau_x = 0, \qquad c\tau_{uu} = 0, \qquad c\xi_u = 0, \qquad c\tau_u = 0, \\
&D_t^\alpha\varphi - uD_t^\alpha\varphi_u = 0, \\
&\mathcal{D}_t^n\xi = 0, \qquad n \in \mathbb{N}, \\
&\binom{\alpha}{n+1}\mathcal{D}_t^{n+1}\tau - \binom{\alpha}{n}D_t^n\varphi_u = 0, \quad n \in \mathbb{N}.
\end{aligned} \tag{4.14}$$

Now, we provide the details of the nonclassical determining equations for Eq. (4.12) using (4.7) and (4.9). To obtain the determining equations we reconsider (4.13) and invariant surface condition $\xi u_x + \tau u_t - \varphi = 0$.
We can distinguish two different cases: $\xi \neq 0$ and $\xi = 0$.
In the case $\xi \neq 0$, without loss of generality, we may assume that $\xi = 1$ and so we have the two situations: $\tau = 0$ and $\tau_u = 0$.
If $\tau = 0$, then $u_x = \varphi$ and differentiating with respect to x yields

$$\begin{aligned}
u_{xx} &= \varphi_x + \varphi\varphi_u, \\
u_{xxx} &= \varphi_{xx} + 2\varphi\varphi_{xu} + \varphi_x\varphi_u + \varphi_u^2\varphi + \varphi^2\varphi_{uu}.
\end{aligned}$$

Substituting these results into (4.13), we get a reduced set of determining equations

$$\begin{aligned}
&a\varphi^2 + a\varphi_x u + b\varphi_{xx} + 2b\varphi\varphi_{xu} + b\varphi^2\varphi_{uu} + 3c\varphi\varphi_{xxu} \\
&+ c\varphi_{xxx} + 3c\varphi_x\varphi_{xu} + 3c\varphi\varphi_u\varphi_{xu} + 3c\varphi^2\varphi_{xuu} + c\varphi^3\varphi_{uuu} \\
&+ 3c\varphi\varphi_x\varphi_{uu} + 3c\varphi^2\varphi_u\varphi_{uu} + d[\varphi - (\varphi_u + 2\varphi)u + \varphi_u u^2] \\
&+ D_t^\alpha\varphi - uD_t^\alpha\varphi_u + \sum_{n=1}^{\infty}\binom{\alpha}{n}D_t^n\varphi_u D_t^{\alpha-n}u + \mu = 0.
\end{aligned} \tag{4.15}$$

If $\tau \neq 0$ and $\tau_u = 0$, hence $u_x = \varphi - \tau u_t$. Differentiating w.r.t. t and x, we see that

$$u_{xt} = \varphi_t + (\varphi_u - \tau_t)u_t - \tau u_{tt},$$

$$u_{xtt} = \varphi_{tt} + (2\varphi_{tu} - \tau_{tt})u_t + \varphi_{uu}u_t^2 + (\varphi_u - 2\tau_t)u_{tt} - \tau u_{ttt},$$

$$u_{xx} = \varphi_x + \varphi\varphi_u - \tau\varphi_t - (2\tau\varphi_u + \tau_x - \tau\tau_t)u_t + \tau^2 u_{tt},$$

$$\begin{aligned} u_{xxt} &= \varphi_{xt} + \varphi\varphi_{tu} + \varphi_t\varphi_u - \tau_t\varphi_t - \tau\varphi_{tt} + [\varphi_{xu} + \varphi_u^2 \\ &+ \varphi\varphi_{uu} - 3\tau\varphi_{tu} - 2\tau_t\varphi_u - \tau_{xt} + \tau_t^2 + \tau\tau_{tt}]u_t - 2\tau\varphi_{uu}u_t^2 \\ &+ [-2\tau\varphi_u - \tau_x + 3\tau\tau_t]u_{tt} + \tau^2 u_{ttt}, \end{aligned}$$

$$\begin{aligned} u_{xxx} &= \varphi_{xx} + 2\varphi\varphi_{xu} + \varphi_x\varphi_u + \varphi\varphi_u^2 + \varphi^2\varphi_{uu} \\ &- 2\tau\varphi_t\varphi_u - \tau\varphi\varphi_{tu} - \tau\varphi_{xt} - 2\tau_x\varphi_t + \tau\tau_t\varphi_t + \tau^2\varphi_{tt} \\ &+ [-3\tau\varphi_{xu} - 3\tau\varphi_u^2 - 3\tau\varphi\varphi_{uu} + 3\tau^2\varphi_{tu} - 3\tau_x\varphi_u - \tau_{xx} \\ &+ 2\tau_x\tau_t + \tau\tau_{xt} + 3\tau\tau_t\varphi_u - \tau\tau_t^2 - \tau^2\tau_{tt}]u_t + 3\tau^2\varphi_{uu}u_t^2 \\ &+ [3\tau^2\varphi_u + 3\tau\tau_x - 3\tau^2\tau_t]u_{tt} - \tau^3 u_{ttt}. \end{aligned}$$

We then substitute these expressions into (4.13) and obtain

$$\begin{aligned}
&a\varphi^2 + a\varphi_x u + a\alpha\tau_t\varphi u + b\varphi_{xx} + 2b\varphi\varphi_{xu} + b\varphi^2\varphi_{uu}\\
&+ b\alpha\tau_t\varphi_x + b\alpha\tau_t\varphi\varphi_u - b\alpha\tau\tau_t\varphi_t - 2b\tau_x\varphi_t + 3c\varphi\varphi_{xxu}\\
&+ c\varphi_{xxx} + 3c\varphi^2\varphi_{xuu} + c\varphi^3\varphi_{uuu} + 3c\varphi_x\varphi_{xu} + 3c\varphi\varphi_u\varphi_{xu}\\
&- 3c\tau\varphi_t\varphi_{xu} + 3c\varphi\varphi_x\varphi_{uu} + 3c\varphi^2\varphi_u\varphi_{uu} - 3c\tau\varphi\varphi_t\varphi_{uu}\\
&+ 2c\alpha\tau_t\varphi\varphi_{xu} + c\alpha\tau_t\varphi_x\varphi_u + c\alpha\tau_t\varphi^2\varphi_{uu} - 2c\alpha\tau\tau_t\varphi_t\varphi_u\\
&- 3c\tau_{xx}\varphi_t + c\alpha\tau_t\varphi_{xx} + c\alpha\tau_t\varphi\varphi_u^2 + c\alpha\tau\tau_t^2\varphi_t - 3c\tau_x\varphi_{xt}\\
&- c\alpha\tau\tau_t\varphi\varphi_{tu} - c\alpha\tau\tau_t\varphi_{xt} - 2c\alpha\tau_t\tau_x\varphi_t + c\alpha\tau^2\tau_t\varphi_{tt}\\
&- 3c\tau_x\varphi\varphi_{tu} - 3c\tau_x\varphi_t\varphi_u + 3c\tau_t\tau_x\varphi_t + 3c\tau\tau_x\varphi_{tt}\\
&+ d[\varphi - (\varphi_u + 2\varphi - \alpha\tau_t)u + (\varphi_u - \alpha\tau_t)u^2] + [-a\tau\varphi\\
&- a\alpha\tau\tau_t u - a\tau_x u - 2b\tau\varphi_{xu} - 2b\tau\varphi\varphi_{uu} - 2b\alpha\tau\tau_t\varphi_u\\
&- b\tau_{xx} - b\alpha\tau_x\tau_t + b\alpha\tau\tau_t^2 - 2b\tau_x\varphi_u + 2b\tau_t\tau_x - c\tau_{xxx}\\
&- 3c\tau\varphi_{xxu} - 6c\tau\varphi\varphi_{xuu} - 3c\tau\varphi^2\varphi_{uuu} - 6c\tau\varphi\varphi_u\varphi_{uu}\\
&+ 3c\tau\tau_t\varphi\varphi_{uu} + 3c\tau\tau_t\varphi_{xu} - 3c\tau_x\varphi\varphi_{uu} - 6c\tau\varphi_u\varphi_{xu}\\
&- 3c\tau_x\varphi_{xu} - 3c\tau\varphi_x\varphi_{uu} - 3c\tau\varphi\varphi_u\varphi_{uu} - 3c\alpha\tau\tau_t\varphi_{xu}\\
&- 3c\alpha\tau\tau_t\varphi\varphi_{uu} - 3c\tau_{xx}\varphi_u + 3c\tau_t\tau_{xx} - 3c\alpha\tau\tau_t\varphi_u^2\\
&+ 3c\tau^2\varphi_t\varphi_{uu} + 3c\alpha\tau^2\tau_t\varphi_{tu} - 3c\alpha\tau_t\tau_x\varphi_u - c\alpha\tau_t\tau_{xx}\\
&+ 2c\alpha\tau_t^2\tau_x + c\alpha\tau\tau_t\tau_{xt} - 3c\tau_x\varphi_{xu} - 3c\tau_x\varphi_u^2 - c\alpha\tau\tau_t^3\\
&- c\alpha\tau^2\tau_t\tau_{tt} + 3c\alpha\tau\tau_t^2\varphi_u - 3c\tau_x\varphi\varphi_{uu} + 9c\tau\tau_x\varphi_{tu}\\
&+ 6c\tau_t\tau_x\varphi_u + 3c\tau_x\tau_{xt} - 3c\tau_t^2\tau_x - 3c\tau\tau_x\tau_{tt}]u_t +\\
&+ [b\tau^2\varphi_{uu} + 3c\tau^2\varphi_{xuu} + 3c\tau^2\varphi\varphi_{uuu} + 6c\tau^2\varphi_u\varphi_{uu}\\
&+ 3c\tau\tau_x\varphi_{uu} - 3c\tau^2\tau_t\varphi_{uu} + 3c\alpha\tau^2\tau_t\varphi_{uu} + 6c\tau\tau_x\varphi_{uu}]u_t^2\\
&- c\tau^3\varphi_{uuu}u_t^3 + [b\alpha\tau^2\tau_t + 2b\tau\tau_x + 3c\tau^2\varphi_{xu} + 3c\tau^2\varphi\varphi_{uu}\\
&+ 3c\tau\tau_{xx} + 3c\alpha\tau^2\tau_t\varphi_u + 3c\alpha\tau\tau_t\tau_x - 3c\alpha\tau^2\tau_t^2 + 3c\tau_x^2\\
&+ 6c\tau\tau_x\varphi_u - 9c\tau\tau_t\tau_x]u_{tt} - 3c\tau^3\varphi_{uu}u_t u_{tt} - [c\alpha\tau^3\tau_t\\
&+ 3c\tau^2\tau_x]u_{ttt} + D_t^\alpha\varphi - uD_t^\alpha\varphi_u + \mu\\
&+ \sum_{n=1}^{\infty}[\binom{\alpha}{n}D_t^n\varphi_u - \binom{\alpha}{n+1}D_t^{n+1}\tau]D_t^{\alpha-n}u = 0.
\end{aligned} \tag{4.16}$$

In the case $\xi = 0$ and $\tau \neq 0$, we obtain the equation $\tau u_t = \varphi$. Differentiating this equation w.r.t. x, we find

$$
\begin{aligned}
u_{xt} &= \frac{1}{\tau^2}[(\tau\varphi_x - \tau_x\varphi) + (\tau\varphi_u - \tau_u\varphi)u_x], \\
u_{xxt} &= \frac{1}{\tau^3}[(\tau^2\varphi_{xx} - \tau\tau_{xx}\varphi - 2\tau\tau_x\varphi_x + 2\tau_x^2\varphi) + (2\tau^2\varphi_{xu} \\
&\quad - 2\tau\tau_{xu}\varphi - 2\tau\tau_x\varphi_u - 2\tau\tau_u\varphi_x + 4\tau_x\tau_u\varphi)u_x + (\tau^2\varphi_{uu} \\
&\quad - \tau\tau_{uu}\varphi - 2\tau\tau_u\varphi_u + 2\tau_u^2\varphi)u_x^2 + (\tau^2\varphi_u - \tau\tau_u\varphi)u_{xx}].
\end{aligned}
$$

Now to get a nonclassical infinitesimal generator, we substitute ξ, u_t, u_{xt} and u_{xxt} into (4.13), which reduces to

$$
\begin{aligned}
&\frac{a}{\tau}(\tau\varphi_x - \tau_x\varphi)u + \frac{b}{\tau^2}(\tau^2\varphi_{xx} - \tau\tau_{xx}\varphi - 2\tau\tau_x\varphi_x + 2\tau_x^2\varphi) \\
&+ \frac{c}{\tau^3}(\tau^3\varphi_{xxx} - \tau^2\tau_{xxx}\varphi - 3\tau^2\tau_{xx}\varphi_x + 6\tau\tau_x\tau_{xx}\varphi - 6\tau_x^3\varphi \\
&+ 6\tau\tau_x^2\varphi_x - 3\tau^2\tau_x\varphi_{xx}) + d\varphi + \frac{d}{\tau}(\alpha\tau\tau_t + \alpha\tau_u\varphi - \tau\varphi_u \\
&- 2\tau\varphi)u - \frac{d}{\tau}(\alpha\tau\tau_t + \alpha\tau_u\varphi - \tau\varphi_u)u^2 + [\frac{a}{\tau}(\tau\varphi + \alpha\tau\tau_t u \\
&+ (\alpha-1)\tau_u\varphi u) + \frac{2b}{\tau^2}(\tau^2\varphi_{xu} - \tau\tau_{xu}\varphi - \tau\tau_x\varphi_u + 2\tau_x\tau_u\varphi \\
&- \tau\tau_u\varphi_x) + \frac{3c}{\tau^3}(\tau^3\varphi_{xxu} - \tau^2\tau_{xxu}\varphi - \tau^2\tau_{xx}\varphi_u - 2\tau^2\tau_{xu}\varphi_x \\
&- 2\tau^2\tau_x\varphi_{xu} - \tau^2\tau_u\varphi_{xx} + 2\tau\tau_u\tau_{xx}\varphi + 4\tau\tau_x\tau_{xu}\varphi + 2\tau\tau_x^2\varphi_u \\
&+ 4\tau\tau_x\tau_u\varphi_x - 6\tau_x^2\tau_u\varphi)]u_x + [\frac{b}{\tau^2}(\tau^2\varphi_{uu} - \tau\tau_{uu}\varphi + 2\tau_u^2\varphi \\
&- 2\tau\tau_u\varphi_u) + \frac{3c}{\tau^3}(\tau^3\varphi_{xuu} - \tau^2\tau_{xuu}\varphi - 2\tau^2\tau_{xu}\varphi_u - \tau^2\tau_x\varphi_{uu} \\
&- \tau^2\tau_{uu}\varphi_x - 2\tau^2\tau_u\varphi_{xu} + 4\tau\tau_u\tau_{xu}\varphi + 2\tau\tau_x\tau_{uu} + 2\tau\tau_u^2\varphi_x\varphi \\
&+ 4\tau\tau_x\tau_u\varphi_u - 6\tau_x\tau_u^2\varphi)]u_x^2 + \frac{c}{\tau^3}[\tau^3\varphi_{uuu} - 3\tau^2\tau_{uu}\varphi_u \\
&- \tau^2\tau_{uuu}\varphi + 6\tau\tau_u\tau_{uu}\varphi - 3\tau^2\tau_u\varphi_{uu} + 6\tau\tau_u^2\varphi_u - 6\tau_u^3\varphi]u_x^3 \\
&+ \frac{1}{\tau^2}[b\alpha\tau^2\tau_t + 6c\tau_x\tau_u\varphi - 3c\tau\tau_u\varphi_x - 3c\tau\tau_x\varphi_u - 3c\tau\tau_{xu}\varphi \\
&+ b(\alpha-1)\tau\tau_u\varphi + 3c\tau^2\varphi_{xu}]u_{xx} + \frac{3c}{\tau^2}[\tau^2\varphi_{uu} - 2\tau\tau_u\varphi_u \\
&- \tau\tau_{uu}\varphi + 2\tau_u^2\varphi]u_xu_{xx} + \frac{c}{\tau}[\alpha\tau\tau_u + (\alpha-1)\tau_u\varphi]u_{xxx} \\
&- \sum_{n=1}^{\infty}[\binom{\alpha}{n+1}\mathcal{D}_t^{n+1}\tau - \binom{\alpha}{n}D_t^n\varphi_u]D_t^{\alpha-n}u \\
&+ D_t^\alpha\varphi - uD_t^\alpha\varphi_u + \mu = 0.
\end{aligned}
\tag{4.17}
$$

Now, we obtain the infinitesimal generators of classical and nonclassical Lie symmetry analysis for fractional diffusion, Burger's, Airy's and Korteweg-de Vries equations; then, we shall discuss the application of these infinitesimal generators.

4.1.1 Fractional diffusion equation

Consider the fractional diffusion equation [with $a = c = d = 0$ and $b \neq 0$ at (4.12)]

$$D_t^\alpha u + bu_{xx} = 0, \quad 0 < \alpha \leq 1, \tag{4.18}$$

where b is constant. After solving the resulting system of determining equations in (4.14) for ξ, τ and φ, we find the infinitesimal generators for the classical symmetry group of Eq. (4.18) to be the following:

$$V_1 = \frac{\alpha}{2} x \frac{\partial}{\partial x} + t \frac{\partial}{\partial t}, \qquad V_2 = \frac{\partial}{\partial x}, \qquad V_3 = u \frac{\partial}{\partial u},$$
$$V_4 = xt^{\alpha-1} \frac{\partial}{\partial u}, \qquad V_5 = t^{\alpha-1} \frac{\partial}{\partial u}.$$

Moreover, equating the coefficients of the various monomials in the first-, second- and fractional-order partial derivatives of u w.r.t t with zero and solving them in (4.15) and (4.16) for ξ, τ and φ, we obtain the infinitesimal generators for the nonclassical symmetry group of Eq. (4.18) as follows:

$$\begin{aligned}
V_6 &= (x + \rho) \frac{\partial}{\partial x} + u \frac{\partial}{\partial u}, \\
V_7 &= (\alpha x + \rho) \frac{\partial}{\partial x} + 2t \frac{\partial}{\partial t}, \\
V_8 &= (\alpha x + \rho) \frac{\partial}{\partial x} + 2t \frac{\partial}{\partial t} + u \frac{\partial}{\partial u}, \\
V_9 &= (\alpha x + \rho) \frac{\partial}{\partial x} + 2t \frac{\partial}{\partial t} + t^{\alpha-1} \frac{\partial}{\partial u}, \\
V_{10} &= (\alpha x + \rho) \frac{\partial}{\partial x} + 2t \frac{\partial}{\partial t} + (\alpha x + \rho) t^{\alpha-1} \frac{\partial}{\partial u},
\end{aligned}$$

where ρ is an arbitrary constant. (Notice that determining Eq. (4.17) is difficult to solve for ξ and φ in general.)
To obtain a classical invariant solution of Eq. (4.18), we consider the one-parameter group generated by $V = V_2 + \beta V_3$ with invariant solution

$$u_1(x, t) = e^{\beta x} f(t),$$

where $\beta \neq 0$ is an arbitrary constant, and substituting this solution into Eq. (4.18), the reduced fractional ordinary differential equation is

$$D_t^\alpha f(t) + b\beta^2 f(t) = 0.$$

The solution of this equation can be written in terms of Mittag–Leffler functions

$$f(t) = t^{\alpha-1}E_{\alpha,\alpha}(-b\beta^2 t^\alpha),$$

thus

$$u_1(x,t) = e^{\beta x}t^{\alpha-1}E_{\alpha,\alpha}(-b\beta^2 t^\alpha). \tag{4.19}$$

We notice that for $\alpha = 1$, this is exactly the solution of diffusion equation with integer order.

Now if $u = u(x,t)$ is an arbitrary solution of Eq. (4.18), by taking the infinitesimal generator

$$V = \lambda_1 V_8 + \lambda_2 V_9 + \lambda_3 V_{10},$$

with $\rho = 0$ and $\lambda_1 + \lambda_2 + \lambda_3 = 1$, we deduce

$$V = \alpha x\frac{\partial}{\partial x} + 2t\frac{\partial}{\partial t} + (\lambda_1 u + \lambda_2 t^{\alpha-1} + \lambda_3 \alpha x t^{\alpha-1})\frac{\partial}{\partial u}.$$

Then, the one-parameter group of transformations G generated by V is given as follows:

$$\begin{aligned}
\bar{x} &= e^{\alpha\varepsilon}x,\\
\bar{t} &= e^{2\varepsilon}t,\\
\bar{u} &= e^{\lambda_1\varepsilon}u + \frac{\lambda_2}{\lambda_1+2-2\alpha}(e^{\lambda_1\varepsilon} - e^{(2\alpha-2)\varepsilon})t^{\alpha-1}\\
&+ \frac{\lambda_3\alpha}{\lambda_1+2-3\alpha}(e^{\lambda_1\varepsilon} - e^{(3\alpha-2)\varepsilon})xt^{\alpha-1},
\end{aligned}$$

and we immediately conclude that

$$\begin{aligned}
u_2(x,t) &= \frac{\lambda_2}{\lambda_1+2-2\alpha}(e^{(\lambda_1+2-2\alpha)\varepsilon} - 1)t^{\alpha-1}\\
&+ \frac{\lambda_3\alpha}{\lambda_1+2-3\alpha}(e^{(\lambda_1+2-3\alpha)\varepsilon} - 1)xt^{\alpha-1}\\
&+ e^{\lambda_1\varepsilon}u(e^{-\alpha\varepsilon}x, e^{-2\varepsilon}t)
\end{aligned} \tag{4.20}$$

is a solution too, where ε is the group parameter. We used the fact that for any given solution u of the equation and for the group of transformations G, (4.20) is a solution of the equation. So, by using the exact solution (4.19), we can write another exact solution as follows:

$$\begin{aligned}
u(x,t) &= \frac{\lambda_2}{\lambda_1+2-2\alpha}[(\frac{\beta}{c})^{\frac{\lambda_1+2-2\alpha}{\alpha}} - 1]t^{\alpha-1}\\
&+ \frac{\lambda_3\alpha}{\lambda_1+2-3\alpha}[(\frac{\beta}{c})^{\frac{\lambda_1+2-3\alpha}{\alpha}} - 1]xt^{\alpha-1}\\
&+ (\frac{\beta}{c})^{\frac{\lambda_1+2-2\alpha}{\alpha}}e^{cx}t^{\alpha-1}E_{\alpha,\alpha}(-bc^2t^\alpha),
\end{aligned}$$

with $c := \beta e^{-\alpha\varepsilon}$.

We now consider the one-parameter group generated by nonclassical infinitesimal generator $V_8 = (\alpha x + \rho)\frac{\partial}{\partial x} + 2t\frac{\partial}{\partial t} + u\frac{\partial}{\partial u}$, in which ρ is a fixed constant. New variables of this group are

$$y = (\alpha x + \rho)^2 t^{-\alpha}, \qquad \nu = t^{-\frac{1}{2}} u. \tag{4.21}$$

Theorem 23 *The transformation (4.21) reduces the FPDE (4.18) to the FODE of the form*

$$\left(\mathcal{P}_{\frac{1}{\alpha}}^{\frac{3}{2}-\alpha,\alpha}\nu\right)(y) + 2b\alpha^2[\nu'(y) + 2y\nu''(y)] = 0.$$

Proof: We use the definition of the Riemann–Liouville differential operator, transformation (4.21) and $0 < \alpha \leq 1$. This yields

$$D_t^\alpha u = \frac{\partial}{\partial t}\frac{1}{\Gamma(1-\alpha)}\int_0^t (t-s)^{-\alpha} s^{\frac{1}{2}} \nu[(\alpha x + \rho)^2 s^{-\alpha}]ds,$$

the substitution $r = \frac{t}{s}$, and we have

$$\begin{aligned} D_t^\alpha u &= \frac{\partial}{\partial t} t^{-\alpha+\frac{3}{2}}(\mathcal{K}_{\frac{1}{\alpha}}^{\frac{3}{2},1-\alpha}\nu)(y) \\ &= t^{-\alpha+\frac{1}{2}}(-\alpha + \frac{3}{2} - \alpha y\frac{d}{dy})(\mathcal{K}_{\frac{1}{\alpha}}^{\frac{3}{2},1-\alpha}\nu)(y) \\ &= t^{-\alpha+\frac{1}{2}}(\mathcal{P}_{\frac{1}{\alpha}}^{\frac{3}{2}-\alpha,\alpha}\nu)(y). \end{aligned}$$

On the other hand, according to (4.21) we find that

$$u_{xx} = 2\alpha^2 t^{-\alpha+\frac{1}{2}}[\nu'(y) + 2y\nu''(y)],$$

which completes the proof.

At this stage, we recall the Example 1 for the one-parameter group generated by $V_5 = t^{\alpha-1}\partial_u$. We derive

$$u(x,t) = c_1 t^{\alpha-1} - h(x)t^{\alpha-1}E_{\alpha,\alpha}(bkt^\alpha),$$

which is a family of solutions for (4.18) where $h''(x) + kh(x) = 0$ and k is arbitrary constant.

4.1.2 Fractional Burger's equation

Consider the fractional Burger's equation [with $a \neq 0, b \neq 0$ and $c = d = 0$ at (4.12)]

$$D_t^\alpha u + auu_x + bu_{xx} = 0, \quad 0 < \alpha \leq 1, \tag{4.22}$$

where a, b are arbitrary constants. This equation appears in a wide variety of physical applications.
Classical generators of Eq. (4.22) can be found as

$$V_1 = \frac{\alpha}{2}x\frac{\partial}{\partial x} + t\frac{\partial}{\partial t} - \frac{\alpha}{2}u\frac{\partial}{\partial u}, \quad V_2 = \frac{\partial}{\partial x};$$

also isolating the coefficients of the various monomials in the first-, second- and fractional-order partial derivatives of u w.r.t t in (4.15) and (4.16) and solving them for ξ, τ and φ, we obtain the nonclassical infinitesimal symmetries of (4.22) as follows:

$$\begin{aligned} V_3 =& (x+\rho)\frac{\partial}{\partial x} + u\frac{\partial}{\partial u}, \\ V_4 =& (\alpha x+\rho)\frac{\partial}{\partial x} + 2t\frac{\partial}{\partial t} - \alpha u\frac{\partial}{\partial u}, \end{aligned}$$

where ρ is arbitrary constant.
Using the nonclassical infinitesimal generator V_3 with invariant solution $u(x,t) = (x+\rho)f(t)$, in this case Eq. (4.22) is reduced to the following fractional ordinary differential equation

$$D_t^{\alpha} f(t) + af^2(t) = 0;$$

the solution of this equation has the form

$$f(t) = -\frac{\Gamma(1-\alpha)}{a\Gamma(1-2\alpha)}t^{-\alpha}.$$

Then the invariant solution of Eq. (4.22) is

$$u(x,t) = -\frac{\Gamma(1-\alpha)}{a\Gamma(1-2\alpha)}(x+\rho)t^{-\alpha}.$$

We now consider the nonclassical infinitesimal generator V_4, and by the change of variables

$$y = (\alpha x + \rho)t^{-\frac{\alpha}{2}}, \quad \nu = t^{\frac{\alpha}{2}}u,$$

Eq. (4.22) is transformed to the fractional ordinary differential equation

$$\left(\mathcal{P}_{\frac{2}{\alpha}}^{1-\frac{3}{2}\alpha,\alpha}\nu\right)(y) + a\alpha\nu(y)\nu'(y) + b\alpha^2\nu''(y) = 0.$$

4.1.3 Fractional Airy's equation

Consider the fractional Airy's equation [with $a = b = d = 0$ and $c \neq 0$ at (4.12)]

$$D_t^{\alpha}u + cu_{xxx} = 0, \quad 0 < \alpha \leq 1, \tag{4.23}$$

where c is fixed constant. Solving overdetermined system of (4.14) in ξ, τ and φ, we find that the Eq. (4.23) admits a six-dimensional classical Lie symmetry algebra spanned by the following generators:

$$\begin{aligned} V_1 =& \frac{\alpha}{3}x\frac{\partial}{\partial x} + t\frac{\partial}{\partial t}, \quad V_2 = \frac{\partial}{\partial x}, \quad V_3 = u\frac{\partial}{\partial u}, \\ V_4 =& x^2 t^{\alpha-1}\frac{\partial}{\partial u}, \quad V_5 = xt^{\alpha-1}\frac{\partial}{\partial u}, \quad V_6 = t^{\alpha-1}\frac{\partial}{\partial u}, \end{aligned}$$

and equating the coefficients of the various monomials in partial derivatives of u w.r.t t and various powers of u in (4.15) and (4.16), we can find the determining equations for the symmetry group of Eq. (4.23). Solving these equations, we obtain the nonclassical infinitesimal symmetries of (4.23) as follows

$$\begin{aligned} V_7 =& (x+\rho)\frac{\partial}{\partial x} + u\frac{\partial}{\partial u}, \\ V_8 =& (\alpha x+\rho)\frac{\partial}{\partial x} + 3t\frac{\partial}{\partial t}, \\ V_9 =& (\alpha x+\rho)\frac{\partial}{\partial x} + 3t\frac{\partial}{\partial t} + u\frac{\partial}{\partial u}, \\ V_{10} =& (\alpha x+\rho)\frac{\partial}{\partial x} + 3t\frac{\partial}{\partial t} + t^{\alpha-1}\frac{\partial}{\partial u}, \\ V_{11} =& (\alpha x+\rho)\frac{\partial}{\partial x} + 3t\frac{\partial}{\partial t} + (\alpha x+\rho)t^{\alpha-1}\frac{\partial}{\partial u}, \\ V_{12} =& (\alpha x+\rho)\frac{\partial}{\partial x} + 3t\frac{\partial}{\partial t} + x(\alpha x+\rho)t^{\alpha-1}\frac{\partial}{\partial u}, \end{aligned}$$

where ρ is arbitrary constant.
Classical vector field $V = V_2 + V_3 + c_1V_4 + c_2V_5 + c_3V_6$ with invariant solution

$$\begin{aligned} u(x,t) =& -\left[c_1x^2 + (2c_1+c_2)x \right.\\ & \left. + 2c_1 + c_2 + c_3\right]t^{\alpha-1} - \mathrm{e}^x f(t) \end{aligned}$$

reduces Eq. (4.23) to the following fractional ordinary differential equation

$$D_t^\alpha f(t) + cf(t) = 0,$$

with solution

$$f(t) = t^{\alpha-1}E_{\alpha,\alpha}(-ct^\alpha).$$

Thus

$$\begin{aligned} u(x,t) =& -\left[c_1x^2 + (2c_1+c_2)x + 2c_1 + c_2 + c_3\right]t^{\alpha-1} \\ & - \mathrm{e}^x t^{\alpha-1}E_{\alpha,\alpha}(-ct^\alpha) \end{aligned}$$

is a solution of the Eq. (4.23).
Now, by using the nonclassical infinitesimal generator V_{12}, and the change of variables

$$y = (\alpha x+\rho)t^{-\frac{\alpha}{3}}, \quad \nu = u - \frac{x^2}{2}t^{\alpha-1},$$

Eq. (4.23) is reduced to the fractional ordinary differential equation as follows:

$$(\mathcal{P}_{\frac{3}{\alpha}}^{1-\alpha,\alpha}\nu)(y) + c\alpha^3\nu'''(y) = 0.$$

4.1.4 Fractional KdV equation

Consider the fractional KdV equation [with $a \neq 0, c \neq 0$ and $b = d = 0$ at (4.12)]

$$D_t^\alpha u + auu_x + cu_{xxx} = 0, \quad 0 < \alpha \leq 1, \tag{4.24}$$

where a, c are arbitrary constants. This equation arises in the theory of long waves in shallow water and other physical systems in which both nonlinear and dispersive effects are relevant.
To obtain the classical infinitesimal symmetries of (4.24), solving the overdetermined system of differential equations in (4.14) for ξ, τ, φ, we find that

$$V_1 = \frac{\alpha}{3}x\frac{\partial}{\partial x} + t\frac{\partial}{\partial t} - \frac{2\alpha}{3}u\frac{\partial}{\partial u}, \quad V_2 = \frac{\partial}{\partial x};$$

also isolating the coefficients of the various monomials in the first-, second- and fractional-order partial derivatives of u w.r.t t in (4.15) and (4.16) and solving them for ξ, τ and φ, we obtain the nonclassical infinitesimal symmetries of (4.24)

$$V_3 = (x+\rho)\frac{\partial}{\partial x} + u\frac{\partial}{\partial u},$$
$$V_4 = (\alpha x+\rho)\frac{\partial}{\partial x} + 3t\frac{\partial}{\partial t} - 2\alpha u\frac{\partial}{\partial u},$$

where ρ is arbitrary constant. Considering the nonclassical infinitesimal generator V_3, the invariant solution $u(x,t) = (x+\rho)f(t)$ reduces Eq. (4.24) to fractional ordinary differential equation

$$D_t^\alpha f(t) + af^2(t) = 0.$$

Then the invariant solution of Eq. (4.24) is

$$u(x,t) = -\frac{\Gamma(1-\alpha)}{a\Gamma(1-2\alpha)}(x+\rho)t^{-\alpha}.$$

By taking the nonclassical infinitesimal generator V_4 and the change of variables

$$y = (\alpha x+\rho)t^{-\frac{\alpha}{2}}, \quad \nu = t^{\frac{\alpha}{2}}u,$$

Eq. (4.22) is reduced to the fractional ordinary differential equation

$$\left(\mathcal{P}_{\frac{3}{\alpha}}^{1-\frac{5}{3}\alpha,\alpha}\nu\right)(y) + a\alpha\nu(y)\nu'(y) + c\alpha^3\nu'''(y) = 0.$$

4.1.5 Fractional gas dynamic equation

Consider the fractional gas dynamic equation [with $b = c = 0$ at (4.12)]

$$D_t^\alpha u + auu_x + du(1-u) = 0, \quad 0 < \alpha \leq 1, \tag{4.25}$$

where $a \neq 0, d \neq 0$ are arbitrary constants. Solving determining equations in (4.13) for ξ, τ and φ, the fractional gas dynamic equation (4.25) is known to admit a three-dimensional classical symmetry group of transformations, with infinitesimal generators

$$V_1 = \frac{\partial}{\partial x}, \quad V_2 = \mathrm{e}^{\frac{d}{a}x}(a\frac{\partial}{\partial x} + du\frac{\partial}{\partial u}).$$

After equating the coefficients of the various monomials in the first-, second- and fractional-order partial derivatives of u w.r.t x and t with zero and solving the resulting system of determining equations in (4.15) and (4.16) for ξ, τ and φ, we find the infinitesimal generators for the nonclassical symmetry group of Eq. (4.25) to be the following:

$$V_3 = \left(a - \rho \mathrm{e}^{\frac{-d}{a}x}\right)\frac{\partial}{\partial x} + du\frac{\partial}{\partial u},$$
$$V_4 = \left(a\rho - \mathrm{e}^{\frac{-d}{a}x}\right)\frac{\partial}{\partial x} + d\rho u\frac{\partial}{\partial u},$$

where ρ is arbitrary constant.
Eq. (4.25) is invariant under the group of transformations generated by classical infinitesimal generator $V_2 = \mathrm{e}^{\frac{d}{a}x}(a\frac{\partial}{\partial x} + du\frac{\partial}{\partial u})$ with invariant solution

$$u = \mathrm{e}^{\frac{d}{a}x}f(t).$$

We find the reduced fractional ordinary differential equation as follows:

$$D_t^\alpha f(t) + df(t) = 0.$$

The solution of this equation is

$$f(t) = t^{\alpha-1}E_{\alpha,\alpha}(-dt^\alpha),$$

and, thus, the exact solution of Eq. (4.25) is

$$u(x,t) = \mathrm{e}^{\frac{d}{a}x}t^{\alpha-1}E_{\alpha,\alpha}(-dt^\alpha). \tag{4.26}$$

Employing the nonclassical infinitesimal generator V_3, invariant solution has the form $u(x,t) = (a\mathrm{e}^{\frac{d}{a}x} - \rho)f(t)$ and Eq. (4.25) is converted to the following fractional ordinary differential equation

$$D_t^\alpha f(t) + df(t) + \rho df^2(t) = 0.$$

4.2 Lie symmetries of space fractional diffusion equations

In this section we consider the space fractional diffusion equation

$$\Delta_1 : u_t - c^2 D_x^\alpha u = 0, \qquad 1 < \alpha \leq 2, \tag{4.27}$$

and seek invariance under the Lie group of transformations given in (4.2). Employing Theorem 22, invariance is given by

$$Pr^{(\alpha,x)}V(\Delta_1) = 0,$$

whenever $\Delta_1 = 0$ where

$$Pr^{(\alpha,x)}V = V + \varphi^t \frac{\partial}{\partial u_t} + \varphi^x \frac{\partial}{\partial u_x} + \varphi^{xx} \frac{\partial}{\partial u_{xx}} + \cdots + \varphi^{(\alpha,x)} \frac{\partial}{\partial D_x^\alpha u}.$$

Now we wish to determine all possible coefficient functions ξ, τ and φ so that the corresponding one-parameter group is a symmetry group of (4.27). Applying $Pr^{(\alpha,x)}V$ to Eq.(4.27), we find infinitesimal criterion

$$\varphi^t - c^2 \varphi^{(\alpha,x)} = 0, \tag{4.28}$$

which must be satisfied whenever Eq.(4.27) holds. We substitute the formulas φ^t and $\varphi^{(\alpha,x)}$ into (4.28) and obtain

$$\begin{aligned} &\varphi_t + (\varphi_u - \tau_t)u_t - \tau_u u_t^2 - \xi_t u_x - \xi_u u_x u_t - c^2 D_x^\alpha \varphi - c^2 \varphi_u D_x^\alpha u \\ &+ c^2 u D_x^\alpha \varphi_u + c^2 \sum_{n=1}^{\infty} \binom{\alpha}{n} \mathcal{D}_x^n \tau D_x^{\alpha-n} u_t + c^2 \alpha \mathcal{D}_x \xi D_x^\alpha u - c^2 \mu \\ &+ c^2 \sum_{n=1}^{\infty} [\binom{\alpha}{n+1} \mathcal{D}_x^{n+1} \xi - \binom{\alpha}{n} D_x^n \varphi_u] D_x^{\alpha-n} u = 0. \end{aligned} \tag{4.29}$$

By replacing $c^2 D_x^\alpha u$ with u_t, we deduce

$$\begin{aligned} &\varphi_t + (\alpha \xi_x - \tau_t)u_t - \tau_u u_t^2 - \xi_t u_x + (\alpha - 1)\xi_u u_x u_t + c^2(u D_x^\alpha \varphi_u - D_x^\alpha \varphi) \\ &+ c^2 \sum_{n=1}^{\infty} \binom{\alpha}{n} \mathcal{D}_x^n \tau D_x^{\alpha-n} u_t - c^2 \mu + c^2 \sum_{n=1}^{\infty} \left[\binom{\alpha}{n+1} \mathcal{D}_x^{n+1} \xi - \binom{\alpha}{n} D_x^n \varphi_u\right] \\ &\times D_x^{\alpha-n} u = 0. \end{aligned} \tag{4.30}$$

We find the determining equations for the symmetry group of Eq. (4.27) as follows:

$$
\begin{aligned}
&\varphi_t = 0, \tau_u = 0, \ \xi_t = 0, \ \ \xi_u = 0,\\
&\alpha\xi_x - \tau_t = 0,\\
&uD_x^{\alpha}\varphi_u - D_x^{\alpha}\varphi = 0,\\
&\mathcal{D}_x^n\tau = 0, \qquad n = 1, 2, 3, ...\\
&\binom{\alpha}{n+1}\mathcal{D}_x^{n+1}\xi - \binom{\alpha}{n}D_x^n\varphi_u = 0. \qquad n = 1, 2, 3, ...
\end{aligned}
$$

We conclude that the most general infinitesimal symmetry of Eq. (4.27) has coefficient functions of the form

$$
\xi = xc_1, \qquad \tau = \alpha t c_1 + c_2, \qquad \varphi = uc_3 + x^{\alpha-2}(xc_4 + c_5),
$$

where $c_1, \cdots, c_5$ are arbitrary constants. Thus the Lie algebra of infinitesimal symmetries of Eq. (4.27) is spanned by the five vector fields

$$
\begin{aligned}
&V_1 = x\frac{\partial}{\partial x} + \alpha t\frac{\partial}{\partial t}, \qquad V_2 = \frac{\partial}{\partial t}, \qquad V_3 = u\frac{\partial}{\partial u},\\
&V_4 = x^{\alpha-1}\frac{\partial}{\partial u}, \qquad V_5 = x^{\alpha-2}\frac{\partial}{\partial u}.
\end{aligned} \tag{4.31}
$$

4.2.1 Nonclassical method

Let (4.27) be a space fractional diffusion equation defined over $M \subset X \times U \simeq R^3$. Let G be a local group of transformations acting on M as in (4.2). To derive the solutions $u = u(t, x)$ of the equation (4.27) such that they are invariant under group transformation (4.2), we have to set

$$
\Delta_2 : \xi u_x + \tau u_t - \varphi = 0.
$$

For the nonclassical method, we seek invariance of both the original equations together with the invariant surface condition. Then we have

$$
Pr^{(\alpha)}V(\Delta_1)|_{\Delta_1=0,\Delta_2=0} = 0, \tag{4.32}
$$

$$
Pr^{(\alpha)}V(\Delta_2)|_{\Delta_1=0,\Delta_2=0} = 0. \tag{4.33}
$$

In this case for every ξ, τ and φ the Eq. (4.32) is identically satisfied when $\Delta_2 = 0$. Indeed, we only consider the condition

$$
\Delta_2 : \xi(x, t, u)u_x + \tau(x, t, u)u_t = \varphi(x, t, u) \tag{4.34}
$$

along with the Eq. (4.33). We can recognize two cases: $\tau \neq 0$ and $\tau = 0$. If $\tau \neq 0$, without loss of generality, we can put $\tau = 1$; thus the equation (4.30) (with replacing τ) implies

$$
\begin{aligned}
&\varphi_t + \varphi_u u_t - \xi_t u_x - \xi_u u_x u_t - c^2D_x^{\alpha}\varphi - c^2\varphi_u D_x^{\alpha}u + c^2uD_x^{\alpha}\varphi_u + c^2\alpha\mathcal{D}_x\xi D_x^{\alpha}u\\
&- c^2\mu + c^2\sum_{n=1}^{\infty}[\binom{\alpha}{n+1}\mathcal{D}_x^{n+1}\xi - \binom{\alpha}{n}D_x^n\varphi_u]D_x^{\alpha-n}u = 0.
\end{aligned}
$$

Substituting the invariant surface condition $u_t = \varphi - \xi u_x$ and also $c^2 D_x^\alpha u = \varphi - \xi u_x$ in the upper equation, we have

$$\varphi_t + \alpha\xi_x\varphi - [\xi_t - (\alpha-1)\xi_u\varphi + \alpha\xi\xi_x]u_x - (\alpha-1)\xi\xi_u u_x^2 + c^2[uD_x^\alpha\varphi_u - D_x^\alpha\varphi]$$
$$-c^2\mu + c^2\sum_{n=1}^{\infty}[\binom{\alpha}{n+1}\mathcal{D}_x^{n+1}\xi - \binom{\alpha}{n}D_x^n\varphi_u]D_x^{\alpha-n}u = 0.$$

We show the determining equations for the nonclassical symmetry group of Eq. (4.27) to be the following:

$$\begin{aligned}
&\varphi_t + \alpha\xi_x\varphi = 0,\\
&\xi_t - (\alpha-1)\xi_u\varphi + \alpha\xi\xi_x = 0,\\
&(\alpha-1)\xi\xi_u = 0,\\
&uD_x^\alpha\varphi_u - D_x^\alpha\varphi = 0,\\
&\binom{\alpha}{n+1}\mathcal{D}_x^{n+1}\xi - \binom{\alpha}{n}D_x^n\varphi_u = 0, \qquad n = 1,2,3,....
\end{aligned}$$

It is clear that the above relations lead to $\mu = 0$. Then this implies that the infinitesimals are

$$\xi = 0, \qquad \varphi = uc_1 + x^{\alpha-1}c_2 + x^{\alpha-2}c_3,$$

or

$$\xi = \frac{x}{\alpha t + \nu}, \qquad \varphi = \frac{c_1}{\alpha t + \nu}u + \frac{c_2}{\alpha t + \nu}x^{\alpha-1} + \frac{c_3}{\alpha t + \nu}x^{\alpha-2},$$

where $c_1, \cdots, c_3$ and ν are arbitrary constants.
If $\tau = 0$, according to (4.34), we can set $u_x = \frac{\varphi}{\xi}$; by replacing this amount in (4.30) we have

$$\varphi_t - \xi_t(\frac{\varphi}{\xi}) - \xi_u(\frac{\varphi}{\xi})u_t + c^2[uD_x^\alpha\varphi_u - D_x^\alpha\varphi] + \alpha\mathcal{D}_x\xi u_t$$
$$-c^2\mu + c^2\sum_{n=1}^{\infty}[\binom{\alpha}{n+1}\mathcal{D}_x^{n+1}\xi - \binom{\alpha}{n}D_x^n\varphi_u]D_x^{\alpha-n}u = 0.$$

We show the determining equations for the nonclassical symmetry group of the equation (4.27) to be the following:

$$\begin{aligned}
&\xi\varphi_t - \xi_t\varphi = 0,\\
&\alpha\xi\xi_x + (\alpha-1)\xi_u\varphi = 0,\\
&uD_x^\alpha\varphi_u - D_x^\alpha\varphi = 0,\\
&\binom{\alpha}{n+1}\mathcal{D}_x^{n+1}\xi - \binom{\alpha}{n}D_x^n\varphi_u = 0. \qquad n = 1,2,3,...
\end{aligned}$$

The equations are difficult to solve for ξ and φ in general.
Comparing with the infinitesimal generator obtained by classical method as in (4.31), we derive additional vector fields as follows:

$$V_6 = x\frac{\partial}{\partial x} + (\alpha t + \nu)\frac{\partial}{\partial t}, \qquad V_7 = x\frac{\partial}{\partial x} + (\alpha t + \nu)\frac{\partial}{\partial t} + \lambda u\frac{\partial}{\partial u},$$
$$V_8 = x\frac{\partial}{\partial x} + (\alpha t + \nu)\frac{\partial}{\partial t} + \gamma x^{\alpha-1}\frac{\partial}{\partial u}, \quad V_9 = x\frac{\partial}{\partial x} + (\alpha t + \nu)\frac{\partial}{\partial t} + \rho x^{\alpha-2}\frac{\partial}{\partial u},$$

where $\nu, \lambda, \gamma, \rho$ are arbitrary constants.

4.3 Lie symmetries of time-fractional diffusion equation

In this section we derive the symmetries of the time-fractional diffusion equation

$$\Delta_3 : D_t^\beta u - c^2 u_{xx} = 0, \qquad 0 < \beta < 1. \tag{4.35}$$

Employing the same argument given in Theorem 22 for Δ_3, the determining equations for classical symmetry are obtained by

$$Pr^{(\beta,t)}V(\Delta_3) = 0,$$

whenever $\Delta_3 = 0$, where infinitesimal generator V is given by (4.3) and

$$Pr^{(\beta,t)}V = V + \varphi^t\frac{\partial}{\partial u_t} + \varphi^x\frac{\partial}{\partial u_x} + \varphi^{xx}\frac{\partial}{\partial u_{xx}} + \varphi^{(\beta,t)}\frac{\partial}{\partial D_t^\beta u},$$

where

$$\varphi^{(\beta,t)} = \mathcal{D}_t^\beta \varphi + \xi D_t^\beta u_x - \mathcal{D}_t^\beta(\xi u_x) - \mathcal{D}_t^{\beta+1}(\tau u) + \mathcal{D}_t^\beta(\mathcal{D}_t(\tau)u) + \tau D_t^{\beta+1}u.$$

Now we wish to determine all possible coefficient functions ξ, τ and φ so that the corresponding one-parameter group is a symmetry group of (4.35). Applying $Pr^{(\beta,t)}V$ to Eq.(4.35), we find infinitesimal criterion

$$\varphi^{(\beta,t)} - c^2\varphi^{xx} = 0, \tag{4.36}$$

which must be satisfied whenever Eq.(4.35) holds. We substitute the formulas φ^{xx} and $\varphi^{(\beta,t)}$ into (4.36) and we derive

$$D_t^\beta \varphi + \varphi_u D_t^\beta u - u D_t^\beta \varphi_u + \sum_{n=1}^{\infty}[\binom{\beta}{n} D_t^n \varphi_u - \binom{\beta}{n+1}\mathcal{D}_t^{n+1}\tau] D_t^{\beta-n}u$$
$$- \sum_{n=1}^{\infty}\binom{\beta}{n}\mathcal{D}_t^n \xi D_t^{\beta-n}u_x - \beta\mathcal{D}_t\tau D_t^\beta u + \mu - c^2\varphi_{xx} - c^2(2\varphi_{xu} - \xi_{xx})u_x$$
$$+ c^2\tau_{xx}u_t - c^2(\varphi_{uu} - 2\xi_{xu})u_x^2 + 2c^2\tau_{xu}u_x u_t + c^2\xi_{uu}u_x^3 + c^2\tau_{uu}u_x^2 u_t - c^2(\varphi_u$$
$$- 2\xi_x)u_{xx} + 2c^2\tau_x u_{xt} + 3c^2\xi_u u_x u_{xx} + c^2\tau_u u_{xx}u_t + 2c^2\tau_u u_x u_{xt} = 0. \tag{4.37}$$

By replacing $D_t^\beta u$ with $c^2 u_{xx}$, we find the determining equations for the symmetry group of the equation (4.35) to be the following:

$$\begin{aligned}
&\varphi_{xx}=0,\tau_{xx}=0,\tau_{xu}=0,\tau_{uu}=0,\tau_x=0,\tau_u=0,\\
&\xi_{uu}=0,\xi_u=0,\\
&2\varphi_{xu}-\xi_{xx}=0,\varphi_{uu}-2\xi_{xu}=0,\beta\tau_t-2\xi_x=0,\\
&D_t^\beta\varphi-uD_t^\beta\varphi_u=0,\\
&\mathcal{D}_t^n\xi=0,\qquad n=1,2,3,...\\
&\binom{\beta}{n}D_t^n\varphi_u-\binom{\beta}{n+1}\mathcal{D}_t^{n+1}\tau=0,\qquad n=1,2,3,....
\end{aligned}$$

We conclude that the most general infinitesimal symmetry of the equation (4.35) has coefficient functions of the form

$$\xi=\frac{\beta}{2}xc_1+c_2,\qquad \tau=tc_1,\qquad \varphi=uc_3+t^{\beta-1}(xc_4+c_5),$$

where $c_1,\cdots,c_5$ arbitrary constants. We notice that $\tau(x,t,u(x,t))|_{t=0}=0$.

4.3.1 Nonclassical method

To apply the nonclassical method to the time-fractional equation, Δ_3, the following symmetry condition needs to satisfy

$$Pr^{(\beta,t)}V(\Delta_3)=0,$$

whenever $\Delta_2=0,\Delta_3=0$, where Δ_2 is the invariant surface condition.
We can recognize two cases: $\xi\neq 0$ and $\xi=0$. If $\xi\neq 0$, without loss of generality, we can put $\xi=1$, thus $u_x=\varphi-\tau u_t$. Differentiating w.r.t. t and x, we see that

$$u_{xt}=\varphi_t+(\varphi_u-\tau_t)u_t-\tau_u u_t^2-\tau u_{tt},$$

$$u_{xx}=\varphi_x+\varphi\varphi_u-\tau\varphi_t-(2\tau\varphi_u+\tau_x+\tau_u\varphi-\tau\tau_t)u_t+2\tau\tau_u u_t^2+\tau^2u_{tt}.$$

Substituting $\xi=1$ and also u_x, u_{xx} and u_{xt} in Eq. (4.40), we have

$$\begin{aligned}
&D_t^\beta\varphi-uD_t^\beta\varphi_u+\sum_{n=1}^{\infty}[\binom{\beta}{n}D_t^n\varphi_u-\binom{\beta}{n+1}\mathcal{D}_t^{n+1}\tau]D_t^{\beta-n}u+\mu\\
&-c^2[\beta\tau_t\varphi_x+\beta\tau_t\varphi\varphi_u-\beta\tau\tau_t\varphi_t+\varphi_{xx}+2\varphi\varphi_{xu}+\varphi^2\varphi_{uu}-2\tau_x\varphi_t-2\tau_u\varphi\varphi_t]\\
&+c^2[2\beta\tau\tau_t\varphi_u+\beta\tau_t\tau_x+\beta\tau_t\tau_u\varphi-\beta\tau\tau_t^2-\beta\tau_u\varphi_x-\beta\tau_u\varphi\varphi_u+\beta\tau\tau_u\varphi_t+2\tau\varphi_{xu}\\
&+\tau_{xx}+2\tau\varphi\varphi_{uu}+2\tau_{xu}\varphi+\tau_{uu}\varphi^2+2\tau_x\varphi_u-2\tau_t\tau_x+\tau_u\varphi_x+\tau_u\varphi\varphi_u-\tau\tau_u\varphi_t\\
&+2\tau_u\varphi\varphi_u-2\tau_t\tau_u\varphi-2\tau\tau_u\varphi_t]u_t-c^2[2\beta\tau\tau_t\tau_u-2\beta\tau\tau_u\varphi_u-\beta\tau_x\tau_u-\beta\tau_u^2\varphi\\
&+\beta\tau\tau_t\tau_u+\tau^2\varphi_{uu}+2\tau\tau_{xu}+2\tau\tau_{uu}\varphi+2\tau_x\tau_u+2\tau\tau_u\varphi_u+\tau_x\tau_u+\tau_u^2\varphi-\tau\tau_t\tau_u\\
&+2\tau_u^2\varphi+2\tau\tau_u\varphi_u-2\tau\tau_t\tau_u]u_t^2+c^2[\tau^2\tau_{uu}+(4-2\beta)\tau\tau_u^2]u_t^3-c^2[\beta\tau^2\tau_t+2\tau\tau_x\\
&+2\tau\tau_u\varphi]u_{tt}-c^2[(\beta-3)\tau^2\tau_u]u_tu_{tt}=0.
\end{aligned}$$

We can equate the coefficients to zero, with the attendant simplification in the formulae, and show the determining equations for the nonclassic symmetry group of the equation (4.35) to be the following:

- $\tau = 0$

$$\begin{aligned}
&D_t^\beta \varphi - uD_t^\beta \varphi_u = 0,\\
&D_t^n \varphi_u = 0, \qquad\qquad n = 1, 2, 3, ...\\
&\varphi_{xx} + 2\varphi\varphi_{xu} + \varphi^2\varphi_{uu} = 0.
\end{aligned}$$

- $\tau \neq 0$

$$\begin{aligned}
&\tau_u = 0,\\
&D_t^\beta \varphi - uD_t^\beta \varphi_u = 0,\\
&\binom{\beta}{n} D_t^n \varphi_u - \binom{\beta}{n+1} D_t^{n+1}\tau = 0, \qquad\qquad n = 1, 2, 3, ...\\
&\beta\tau_t\varphi_x + \beta\tau_t\varphi\varphi_u + \varphi_{xx} + 2\varphi\varphi_{xu} = 0,\\
&-2\tau_x\varphi_u + \beta\tau_t\tau_x + 2\tau\varphi_{xu} + \tau_{xx} = 0,\\
&\varphi_{uu} = 0,\\
&\beta\tau\tau_t + 2\tau_x = 0.
\end{aligned}$$

Then this implies that the infinitesimals are

- $\tau = 0$

$$\varphi = \frac{u}{x+\nu} + [\frac{c_1}{x+\nu} + c_2(x+\nu)^2]t^{\beta-1},$$

or

$$\varphi = uc_1 + (xc_2 + c_3)t^{\beta-1}.$$

- $\tau \neq 0$

$$\tau = \frac{2t}{\beta x+\nu}, \qquad \varphi = \frac{c_1}{\beta x+\nu}u + [\frac{c_2}{\beta x+\nu} + c_3]t^{\beta-1},$$

where $c_1, \cdots, c_3$ and ν are arbitrary constants.
If $\xi = 0$, thus $u_t = \frac{\varphi}{\tau}$. Differentiating w.r.t. x, we have

$$u_{xt} = \frac{(\tau\varphi_x - \tau_x\varphi) + (\tau\varphi_u - \tau_u\varphi)}{\tau^2}.$$

By replacing this amount in (4.40), we obtain

$$\begin{aligned}
&D_t^\beta\varphi - uD_t^\beta\varphi_u + \sum_{n=1}^{\infty}[\binom{\beta}{n}D_t^n\varphi_u - \binom{\beta}{n+1}D_t^{n+1}\tau]D_t^{\beta-n}u + \mu \\
&+ c^2[\beta\tau\tau_t + (\beta-1)\tau_u\varphi]u_{xx} - c^2(\tau^2\varphi_{xx} - \tau\tau_{xx}\varphi - 2\tau\tau_x\varphi_x + 2\tau_x^2\varphi) \\
&- c^2(2\tau^2\varphi_{xu} - 2\tau\tau_{xu}\varphi - 2\tau\tau_x\varphi_u + 4\tau_x\tau_u\varphi - 2\tau\tau_u\varphi_x)u_x \\
&- c^2(\tau^2\varphi_{uu} - \tau\tau_{uu}\varphi - 2\tau\tau_u\varphi_u + 2\tau_u^2\varphi)u_x^2 = 0.
\end{aligned}$$

We achieve the determining equations for the nonclassic symmetry group of the equation (4.35) to be the following:

$$\begin{aligned}
&D_t^\beta\varphi - uD_t^\beta\varphi_u = 0, \\
&\binom{\beta}{n}D_t^n\varphi_u - \binom{\beta}{n+1}D_t^{n+1}\tau = 0, \qquad n = 1, 2, 3, ... \\
&\beta\tau\tau_t + (\beta-1)\tau_u\varphi = 0, \\
&\tau^2\varphi_{xx} - \tau\tau_{xx}\varphi - 2\tau\tau_x\varphi_x + 2\tau_x^2\varphi = 0, \\
&2\tau^2\varphi_{xu} - 2\tau\tau_{xu}\varphi - 2\tau\tau_x\varphi_u + 4\tau_x\tau_u\varphi - 2\tau\tau_u\varphi_x = 0, \\
&\tau^2\varphi_{uu} - \tau\tau_{uu}\varphi - 2\tau\tau_u\varphi_u + 2\tau_u^2\varphi = 0.
\end{aligned}$$

The equations are difficult to solve for ξ and φ in general.
Infinitesimal generators obtained by the nonclassical method of time-fractional diffusion equation are as follows:

$$\begin{aligned}
&V_1 = (x+\nu)\frac{\partial}{\partial x} + u\frac{\partial}{\partial u}, \qquad V_2 = (x+\nu)\frac{\partial}{\partial x} + (u + \lambda t^{\beta-1})\frac{\partial}{\partial u}, \\
&V_3 = (x+\nu)\frac{\partial}{\partial x} + [u + \rho(x+\nu)^3 t^{\beta-1}]\frac{\partial}{\partial u}, \qquad V_4 = (\beta x+\nu)\frac{\partial}{\partial x} + 2t\frac{\partial}{\partial t}, \\
&V_5 = (\beta x+\nu)\frac{\partial}{\partial x} + 2t\frac{\partial}{\partial t} + u\frac{\partial}{\partial u}, \qquad V_6 = (\beta x+\nu)\frac{\partial}{\partial x} + 2t\frac{\partial}{\partial t} + t^{\beta-1}\frac{\partial}{\partial u}, \\
&V_7 = (\beta x+\nu)\frac{\partial}{\partial x} + 2t\frac{\partial}{\partial t} + (\beta x+\nu)t^{\beta-1}\frac{\partial}{\partial u},
\end{aligned}$$

where λ, ν, ρ are arbitrary constants.

4.4 General solutions to fractional diffusion equations by invariant surfaces

In this section we give an idea of construction of the solutions to an equation using a new invariant condition. For a given invariance to a fractional partial differential equation, reduction to a solvable fractional ordinary differential equation is not straightforward to capture invariant solutions. We are now ready to state the idea of finding the structure of solutions to fractional differential equations for which the invariance has been found. Let the Eq. (4.4) admit the group of transformation (4.2) with infinitesimal generator

$$V = \xi(x,t,u)\frac{\partial}{\partial x} + \tau(x,t,u)\frac{\partial}{\partial t} + \varphi(x,t,u)\frac{\partial}{\partial u}.$$

If the solution $g(x,t,u) = u(x,t) - u = 0$ is invariant, then it satisfies the surface condition 4.34. However, if there exists a class of solutions $g(x,t,u) = c$ to the Eq. (4.4) admitting group (4.2), $g(\bar{x},\bar{t},\bar{u}) = \bar{c}$ defining a solution too, then by Theorem (2.2.7-3) in [28], we have

$$Vg(x,t,u) = 1. \tag{4.38}$$

We merge this equation and the original equation to obtain the invariant class of solutions. For instance we have the following example.

Example 2 *Consider the one-parameter group generated by infinitesimal generator $V = (k_1 x^{\alpha-1} + k_2 x^{\alpha-2})\frac{\partial}{\partial u}$ of the Eq. (4.27), where k_1, k_2 are constants. Then according to (4.38), we can write*

$$(k_1 x^{\alpha-1} + k_2 x^{\alpha-2})g_u = 1.$$

The solution of this equation can be written of the form

$$g(x,t,u) = \frac{u + w(x,t)}{k_1 x^{\alpha-1} + k_2 x^{\alpha-2}} = c_1,$$

so

$$u = a_1 x^{\alpha-1} + a_2 x^{\alpha-2} - w(x,t),$$

where $a_1 := c_1 k_1$, $a_2 := c_1 k_2$. For example if $w(x,t) = f(x)h(t)$, then

$$u(x,t) = a_1 x^{\alpha-1} + a_2 x^{\alpha-2} - f(x)h(t).$$

Substituting this solution into the Eq. (4.27) and letting $h'(t) - c^2 h(t) = 0$, the reduced fractional ordinary differential equation is

$$f(x) - D_x^\alpha f(x) = 0.$$

The solution of this equation can be written in terms of Mittag-Leffler functions

$$f(x) = a_3 x^{\alpha-1} E_{\alpha,\alpha}(x^\alpha);$$

thus

$$u(x,t) = a_1 x^{\alpha-1} + a_2 x^{\alpha-2} + a_3 e^{c^2 t} x^{\alpha-1} E_{\alpha,\alpha}(x^\alpha)$$

is a family of solutions (4.27).

Now, by using some of the vector fields obtained from the classical and nonclassical Lie symmetries, we represent exact solutions of fractional diffusion equations.

Example 3 *Equation* (4.27) *with infinitesimal generator*

$$V = \frac{\partial}{\partial t} + \delta u \frac{\partial}{\partial u}$$

has invariant solution

$$u(x,t) = e^{\delta t} f(x),$$

where δ is an arbitrary constant; substituting this solution into Eq. (4.27), *the reduced fractional ordinary differential equation is*

$$D_x^\alpha f(x) = \frac{\delta}{c^2} f(x).$$

The solution of this equation can be written in terms of Mittag-Leffler functions

$$f(x) = c_1 x^{\alpha-1} E_{\alpha,\alpha}(\frac{\delta}{c^2} x^\alpha),$$

where c_1 is an arbitrary constant; thus

$$u(x,t) = c_1 e^{\delta t} x^{\alpha-1} E_{\alpha,\alpha}(\frac{\delta}{c^2} x^\alpha). \tag{4.39}$$

We notice that for $\alpha = 2$, $u(x,t) = \frac{c}{\sqrt{\lambda}} e^{\lambda t} sinh\left(\frac{\sqrt{\lambda}}{c} x\right)$ is an exact solution of diffusion equation with integer order. Now if $u = u(x,t)$ is an arbitrary solution of the Eq. (4.27), *by taking the nonclassical infinitesimal generator $V = V_7 + V_8 + V_9$ with $\nu = 0$, we deduce the one-parameter group of transformations G generated by V is given as follows:*

$\bar{x} = e^{3\varepsilon} x,$

$\bar{t} = e^{3\alpha\varepsilon} t,$

$\bar{u} = e^{\lambda\varepsilon} u + \frac{\gamma}{3\alpha - 3 - \lambda}(e^{(3\alpha-3)\varepsilon} - e^{\lambda\varepsilon}) x^{\alpha-1} + \frac{\rho}{3\alpha - 6 - \lambda}(e^{(3\alpha-6)\varepsilon} - e^{\lambda\varepsilon}) x^{\alpha-2},$

and we immediately conclude that

$$u_2(x,t) = \frac{\gamma}{\lambda+3-3\alpha}(e^{(\lambda+3-3\alpha)\varepsilon}-1)x^{\alpha-1}$$
$$+\frac{\rho}{\lambda+6-3\alpha}(e^{(\lambda+6-3\alpha)\varepsilon}-1)x^{\alpha-2}+e^{\lambda\varepsilon}u(e^{-3\varepsilon}x,e^{-3\alpha\varepsilon}t) \quad (4.40)$$

is a solution too, where ε *is the group parameter. We used the fact that for any given solution* u *of equation and for the group of transformations* G, (4.40) *is a solution of the equation. So by using the exact solution* (4.39), *we can write another exact solution as follows:*

$$u_2(x,t) = c_2x^{\alpha-1}+c_3x^{\alpha-2}+c_4e^{kt}x^{\alpha-1}E_{\alpha,\alpha}(\frac{k}{c^2}x^{\alpha}),$$

where $c_2 := \frac{\gamma}{\lambda+3-3\alpha}(e^{(\lambda+3-3\alpha)\varepsilon}-1)$, $c_3 := \frac{\rho}{\lambda+6-3\alpha}(e^{(\lambda+6-3\alpha)\varepsilon}-1)$, $c_4 := c_1e^{(\lambda+3-3\alpha)\varepsilon}$ *and* $k := \delta e^{-3\alpha\varepsilon}$.

Example 4 *We now consider the one-parameter group generated by nonclassical infinitesimal generator* $V_7 = x\frac{\partial}{\partial x}+(\alpha t+\nu)\frac{\partial}{\partial t}+\lambda u\frac{\partial}{\partial u}$ *of the Eq.* (4.27), *in which* ν *and* λ *are fixed constants. New variables of this group are*

$$y = (\alpha t+\nu)x^{-\alpha}, \qquad \mathcal{F} = x^{-\lambda}u. \quad (4.41)$$

Theorem 24 *The transformation* (4.41) *reduces the FPDE* (4.27) *to the FODE of the form*

$$\alpha\mathcal{F}'(y)-c^2(\mathcal{P}^{\lambda+1-\alpha,\alpha}_{\frac{1}{\alpha}}\mathcal{F})(y) = 0.$$

Proof: *We use the definition of the Riemann-Liouville differential operator, and* $1<\alpha\leq 2$. *This yields*

$$D^{\alpha}_{x}u = \frac{\partial^2}{\partial x^2}\frac{1}{\Gamma(2-\alpha)}\int_0^x (x-s)^{1-\alpha}s^{\lambda}\mathcal{F}[(\alpha t+\nu)s^{-\alpha}]ds,$$

and using the substitution $r=\frac{x}{s}$, *we have*

$$D^{\alpha}_{x}u = \frac{\partial^2}{\partial x^2}x^{2+\lambda-\alpha}(\mathcal{K}^{\lambda+1,2-\alpha}_{\frac{1}{\alpha}}\mathcal{F})(y)$$
$$= x^{\lambda-\alpha}(\lambda-\alpha+1-\alpha y\frac{d}{dy})(\lambda-\alpha+2-\alpha y\frac{d}{dy})(\mathcal{K}^{\lambda+1,2-\alpha}_{\frac{1}{\alpha}}\mathcal{F})(y)$$
$$= x^{\lambda-\alpha}(\mathcal{P}^{\lambda+1-\alpha,\alpha}_{\frac{1}{\alpha}}\mathcal{F})(y).$$

On the other hand, we find that

$$u_t = \alpha x^{\lambda-\alpha}\mathcal{F}'(y),$$

which completes the proof.

Example 5 *Equation (4.27) is invariant under the group of transformations generated by*

$$V = \frac{\partial}{\partial t} + x^{\alpha-1}\frac{\partial}{\partial u},$$

with invariant solution

$$u(x,t) = x^{\alpha-1}t + f(x).$$

Substituting this solution into Eq. (4.27), we find the reduced fractional ordinary differential equation

$$D_x^\alpha f(x) = \frac{1}{c^2}x^{\alpha-1}.$$

The solution of this equation is

$$f(x) = \frac{\Gamma(\alpha)}{c^2\Gamma(2\alpha)}x^{2\alpha-1} + k_1 x^{\alpha-1} + k_2 x^{\alpha-2},$$

where k_1, k_2 are arbitrary constants; then

$$u(x,t) = x^{\alpha-1}t + \frac{\Gamma(\alpha)}{c^2\Gamma(2\alpha)}x^{2\alpha-1} + k_1 x^{\alpha-1} + k_2 x^{\alpha-2}.$$

Example 6 *Equation (4.35) with infinitesimal generator*

$$V = \frac{\partial}{\partial x} + (\lambda_1 u + \lambda_2 x t^{\beta-1} + \lambda_3 t^{\beta-1})\frac{\partial}{\partial u}$$

has invariant solution

$$u(x,t) = e^{\lambda_1 x}f(t) - \frac{\lambda_2}{\lambda_1}(x + \frac{1}{\lambda_1})t^{\beta-1} - \frac{\lambda_3}{\lambda_1}t^{\beta-1},$$

where $\lambda_1, \cdots, \lambda_3$ are arbitrary constants. Substituting this solution into Eq. (4.35), the reduced fractional ordinary differential equation is

$$D_t^\beta f(t) = \lambda_1^2 c^2 f(t).$$

The solution of this equation can be written in terms of Mittag-Leffler functions

$$f(t) = t^{\beta-1}E_{\beta,\beta}(\lambda_1^2 c^2 t^\beta),$$

thus

$$u(x,t) = e^{\lambda_1 x}t^{\beta-1}E_{\beta,\beta}(\lambda_1^2 c^2 t^\beta) - \frac{\lambda_2}{\lambda_1}(x + \frac{1}{\lambda_1})t^{\beta-1} - \frac{\lambda_3}{\lambda_1}t^{\beta-1}.$$

Example 7 *Equation (4.35) is invariant under the group of transformations generated by*

$$V = (x+\nu)\frac{\partial}{\partial x} + (u + \lambda t^{\beta-1})\frac{\partial}{\partial u}$$

with invariant solution

$$u(x,t) = -\lambda t^{\beta-1} - (x+\nu)f(t),$$

where ν, λ are arbitrary constants. Substituting this solution into Eq. (4.27), we find the reduced fractional ordinary differential equation

$$D_t^{\beta} f(t) = 0.$$

The solution of this equation is

$$f(t) = k_1 t^{\beta-1},$$

where k_1 is an arbitrary constant; then

$$u(x,t) = -\lambda t^{\beta-1} - k_1(x+\nu)t^{\beta-1}.$$

Chapter 5

Conservation laws of the fractional differential equations

A new technique for constructing conservation laws for fractional differential equations not having a Lagrangian is proposed. The technique is based on the methods of Lie group analysis and employs the concept of nonlinear self-adjointness which is enhanced to the certain class of fractional evolution equations. The proposed approach is demonstrated on subdiffusion and diffusion-wave equations with the Riemann–Liouville and Caputo time-fractional derivatives. It is shown that these equations are nonlinearly self-adjoint, and, therefore, desired conservation laws can be calculated using the appropriate formal Lagrangians. The explicit forms of fractional generalizations of the Noether operators are also proposed for the equations with the Riemann-Liouville and Caputo time-fractional derivatives of order $\alpha \in (0, 2)$. Using these operators and formal Lagrangians, new conservation laws are constructed for the linear and nonlinear time-fractional subdiffusion and diffusion-wave equations by their Lie point symmetries. Recently, it was shown in [92, 17] that conservation laws can be also efficiently used for constructing the particular solutions of PDEs and its systems.

In 1996, Riewe [161] introduced a Lagrangian depending on fractional derivatives. During the last two decades, many fractional generalizations of the Euler–Lagrange equations with different types of fractional derivatives were derived [7, 24, 23, 19, 90, 118, 8]. Using these results, several fractional generalizations of Noether's theorem were proved [56, 16, 127, 149, 31], and some number of fractional conservation laws were calculated for equations and systems having different fractional Lagrangians [57, 169].

However, conservation laws still are not widely used for investigation of the properties of fractional differential equations. The main reason is the fact that most fractional partial differential equations such as fractional diffusion and transport equations, fractional kinetic equations and fractional relaxation equations (see, e.g., [132, 164, 112, 124, 111, 20, 185] and references therein) do not have a Lagrangian. Thus, Noether's theorem and its fractional generalizations cannot be used for obtaining conservation laws for such equations.

In this section, a new technique for constructing conservation laws is applied for fractional differential equations not having a Lagrangian in a classical sense. The technique uses the modern methods of Lie group analysis of fractional differential equations developed in [66, 67, 22, 21, 101, 125] and employs the concept of nonlinear self-adjointness proposed for inter-order differential equations. This concept is based on the notion of *formal Lagrangian* and provides to construct conservation laws for nonlinearly self-adjoint differential equations using classical algorithms.

5.1 Description of approach

We describe here the mentioned method by some well-known equations.

5.1.1 Time-fractional diffusion equations

Let us consider a nonlinear time-fractional diffusion equation (TFDE)

$$\mathcal{D}_t^\alpha u = (k(u)u_x)_x, \qquad \alpha \in (0,2). \tag{5.1}$$

Here, u is a function of the independent variables $t \in (0,T](T \leq \infty)$ and $x \in \Omega \subset \mathbb{R}$, $\mathcal{D}_t^\alpha u$ is a fractional derivative of function u with respect to t of order α, and k is a diffusion coefficient which is considered as a function of the dependent variable u.

We consider two different types of fractional derivatives $\mathcal{D}_t^\alpha u$ in Eq. (5.1): one is the Riemann-Liouville left-sided time-fractional derivative ${}_0^{RL}D_t^\alpha u$, and the other is the Caputo left-sided time-fractional derivative ${}_0^{C}D_t^\alpha u$. Equation (5.1) is known as the subdiffusion equation for $\alpha \in (0,1)$ and as the diffusion-wave equation for $\alpha \in (1,2)$.

The symmetry properties of Eq. (5.1) have been investigated in [67], and Lie point symmetries for this equation with the Riemann-Liouville and Caputo time-fractional derivatives have been obtained there. These symmetries will be used here to construct the conservation laws for Eq. (5.1).

5.1.2 Conservation laws and nonlinear self-adjointness

Let us define a conservation law for Eq. (5.1) in the same manner as it is defined for the classical diffusion and wave equations. Namely, a vector field $\mathcal{C} = (\mathcal{C}^t, \mathcal{C}^x)$ where $\mathcal{C}^t = \mathcal{C}^t(t,x,u,...)$, $\mathcal{C}^x = \mathcal{C}^x(t,x,u,...)$ is called a *conserved vector* for Eq. (5.1) if it satisfies the conservation equation

$$D_t\mathcal{C}^t + D_x\mathcal{C}^x = 0, \tag{5.2}$$

on all solutions of Eq. (5.1). Equation (5.2) is called a *conservation law* for Eq. (5.1).

A conserved vector is called a *trivial conserved vector* for Eq. (5.1) if its components $\mathcal{C}^t$ and $\mathcal{C}^x$ vanish on the solution of this equation.

Note that Eq. (5.1) with the Riemann–Liouville fractional derivative can be rewritten in the form of conservation law (5.2) with

$$\mathcal{C}^t = D_t^{n-1}\left({}_0I_t^{n-\alpha}u\right), \quad \mathcal{C}^x = -k(u)u_x, \quad n = 1, 2. \tag{5.3}$$

It is important to point out that the order $n - \alpha$ of fractional integral in (5.3) is the same as the one used in Eqs. (5.1) with (5.2).

In the case of the Caputo fractional derivative, Eq. (5.1) can also be rewritten in the form of conservation law (5.2) with

$$\mathcal{C}^t = {}_0I_t^{n+1-\alpha}\left(D_t^n u\right), \quad \mathcal{C}^x = -k(u)u_x, \quad n = 1, 2. \tag{5.4}$$

Contrary to the previous case, the order of fractional integral in (5.4) has been increased by one. In other words, the coordinate $\mathcal{C}^t$ now depends on a new integral variable which is lacking in Eq. (5.1).
In accordance with the concept of nonlinear self-adjointness [44], a formal Lagrangian for this equation can be introduced as

$$\mathcal{L} = v(t,x)\left[\mathcal{D}_t^\alpha u - k'(u)u_x^2 - k(u)u_{xx}\right], \tag{5.5}$$

where v is a new dependent variable. In view of this formal Lagrangian, an action integral is

$$\int_0^t \int_\Omega \mathcal{L}(t, x, u, v, \mathcal{D}_t^\alpha u, u_x, u_{xx})dxdt. \tag{5.6}$$

Assume that variable v in the action (5.6) is not varied. Then, using fractional variational approach developed by Agrawal [16], one can find the Euler-Lagrange operator with respect to u corresponding to the action (5.6) as

$$\frac{\delta}{\delta u} = \frac{\partial}{\partial u} + \left(\mathcal{D}_t^\alpha u\right)^* \frac{\partial}{\partial \mathcal{D}_t^\alpha u} - D_x\frac{\partial}{\partial u_x} + D_x^2\frac{\partial}{\partial u_{xx}}. \tag{5.7}$$

Here, $(\mathcal{D}_t^\alpha)^*$ is the adjoint operator of $\mathcal{D}_t^\alpha$. The corresponding adjoint operators are

$$\begin{aligned}\left({}_0^{RL}D_t^\alpha\right)^* &= (-1)^n {}_tI_T^{n-\alpha}\left(D_t^n\right) \equiv {}_t^C D_T^\alpha,\\ \left({}_0^C D_t^\alpha\right)^* &= (-1)^n D_t^n\left({}_tI_T^{n-\alpha}\right) \equiv {}_t^{RL}D_T^\alpha.\end{aligned}$$

Here, ${}_tI_T^{n-\alpha}$ is the right-sided operator of fractional integration of order $n - \alpha$ defined by

$$\left({}_0I_T^{n-\alpha}f\right)(t,x) = \frac{1}{\Gamma(n-\alpha)}\int_t^T \frac{f(\tau,x)}{(\tau-t)^{\alpha+1-n}}d\tau.$$

Similar to the case of integer-order nonlinear differential equations, the *adjoint equation* to the nonlinear TFDE (5.1) can be defined as Euler– Lagrange equation

$$\frac{\partial \mathcal{L}}{\partial u} = 0, \tag{5.8}$$

where $\mathcal{L}$ is the formal Lagrangian (5.5) and $\frac{\delta}{\delta u}$ is the Euler–Lagrange operator (5.7). After calculations, Eq. (5.8) takes the form

$$(\mathcal{D}_t^\alpha)^* \, v - k(u)v_{xx} = 0, \quad n = [\alpha] + 1, \quad \alpha \in (0, 2). \tag{5.9}$$

Eq. (5.1) will be called *nonlinearly self-adjoint* if the adjoint Eq. (5.9) is satisfied for all solutions u of Eq. (5.1) upon a substitution $v = \varphi(t, x, u)$ such that $\varphi(t, x, u) \neq 0$.

5.1.3 Fractional Noether operators

It is unwieldy to construct the conserved vectors by direct use of Noether's theorem. A more convenient approach for integer-order differential equations was proposed in [100]. In this approach, the components of conserved vectors are obtained by applying the so-called Noether operators to the Lagrangian. These Noether operators can be found from the *fundamental operator identity*, also known as the Noether identity. For the considered case of two independent variables t, x, and one dependent variable $u(t, x)$, this fundamental identity can be written as

$$\overline{X} + D_t(\xi^1)\mathcal{I} + D_x(\xi^2)\mathcal{I} = \mathcal{W}\frac{\delta}{\delta u} + D_t\mathcal{N}^t + D_x\mathcal{N}^x. \tag{5.10}$$

Here, $\mathcal{I}$ is the identity operator, $\frac{\delta}{\delta u}$ is the Euler–Lagrange operator, $\mathcal{N}^t$ and $\mathcal{N}^x$ are the Noether operators, $\overline{X}$ is an appropriate prolongation for the Lie point group generator

$$X = \xi^1(t, x, u)\frac{\partial}{\partial t} + \xi^2(t, x, u)\frac{\partial}{\partial x} + \phi(t, x, u)\frac{\partial}{\partial u} \tag{5.11}$$

to all derivatives (integer and/or fractional order) of the dependent variable $u(t, x)$ which are contained in the considered equation, and $\mathcal{W} = \phi - \xi^1 u_t - \xi^2 u_x$.

The prolongation of the generator (5.11) for Eq. (5.1) has the form

$$\begin{aligned} \overline{X} =& \xi^1\frac{\partial}{\partial t} + \xi^2\frac{\partial}{\partial x} + \phi\frac{\partial}{\partial u} \\ &+ \zeta_\alpha^0\frac{\partial}{\partial(D_t^\alpha u)} + \zeta_1^1\frac{\partial}{\partial u_x} + \zeta_2^1\frac{\partial}{\partial u_{xx}} \end{aligned} \tag{5.12}$$

where $\zeta_\alpha^0, \zeta_1^1, \zeta_2^1$ are given by the prolongation formulae

$$\zeta_\alpha^0 = \mathcal{D}_t^\alpha(\mathcal{W}) + \xi^1 D_t(\mathcal{D}_t^\alpha u) + \xi^2 D_x(\mathcal{D}_t^\alpha u),$$
$$\zeta_1^1 = D_x(\mathcal{W}) + \xi^1 u_{tx} + \xi^2 u_{xx},$$
$$\zeta_2^1 = D_x^2(\mathcal{W}) + \xi^1 u_{txx} + \xi^2 u_{xxx}.$$

For given operators (5.7) and (5.12), one can verify that the equality (5.10) is fulfilled if the Noether operators are defined as follows. For the case when the Riemann–Liouville time-fractional derivative is used in Eq. (5.1), the operator $\mathcal{N}^t$ is

$$\begin{aligned}\mathcal{N}^t =& \xi^1\mathcal{I} + \sum_{k=0}^{n-1}(-1)^k {}_0D_t^{\alpha-1-k}(\mathcal{W}) D_t^k \frac{\partial}{\partial({}_0^{RL}D_t^\alpha u)} \\ & - (-1)^n \mathfrak{J}\left(\mathcal{W}, D_t^n \frac{\partial}{\partial({}_0^{RL}D_t^\alpha u)}\right).\end{aligned} \tag{5.13}$$

For an other case when the Caputo time-fractional derivative is used in Eq. (5.1), this operator takes the form

$$\begin{aligned}\mathcal{N}^t =& \xi^1\mathcal{I} + \sum_{k=0}^{n-1} D_t^k(\mathcal{W}) {}_tD_T^{\alpha-1-k} \frac{\partial}{\partial({}_0^{C}D_t^\alpha u)} \\ & - \mathfrak{J}\left(D_t^n(\mathcal{W}), \frac{\partial}{\partial({}_0^{C}D_t^\alpha u)}\right).\end{aligned} \tag{5.14}$$

The operator $\mathcal{N}^x$ in both cases is

$$\mathcal{N}^x = \xi^2\mathcal{I} + \mathcal{W}\left(\frac{\partial}{\partial u_x} - D_x\frac{\partial}{\partial u_{xx}}\right) + D_x(\mathcal{W})\frac{\partial}{\partial u_{xx}}. \tag{5.15}$$

In (5.13) and (5.14), $\mathfrak{J}$ is the integral

$$\mathfrak{J}(f,g) = \frac{1}{\Gamma(n-\alpha)}\int_0^t\int_t^T \frac{f(\tau,x)g(\mu,x)}{(\mu-\tau)^{\alpha+1-n}}d\mu d\tau. \tag{5.16}$$

Now assume that Eq. (5.1) is nonlinearly self-adjoint. This means that a function $v = \varphi(t,x,u)$ exists such that Eq. (5.9) is satisfied for any solution of Eq. (5.1). Then, the explicit formulae for the components of conserved vectors associated with the symmetries of Eq. (5.1) can be established.

We act on the formal Lagrangian (5.5) by both sides of the Noether identity (5.10). For any generator X admitted by Eq. (5.1) and any solution of this equation, the left-hand side of this equality is equal to zero:

$$\left(\overline{X}\mathcal{L} + D_t(\xi^1)\mathcal{L} + D_x(\xi^2)\mathcal{L}\right)\Big|_{(5.1)} = 0.$$

For nonlinearly self-adjoint equations, the Euler-Lagrange Eq. (5.8) is identically zero. Therefore, the right-hand side of the equality under consideration leads to the conservation law

$$D_t(\mathcal{N}^t\mathcal{L}) + D_x(\mathcal{N}^x\mathcal{L}) = 0, \tag{5.17}$$

where the operator $\mathcal{N}^t$ is defined by (5.13) or (5.14), and operator $\mathcal{N}^x$ is defined by (5.15).

From the comparison of (5.2) and (5.17), it is easy to conclude that any Lie point symmetry of Eq. (5.1) gives the conserved vector for this equation with components defined by the explicit formulae

$$\mathcal{C}^t = \mathcal{N}^t(\mathcal{L}), \qquad \mathcal{C}^x = \mathcal{N}^x(\mathcal{L}). \tag{5.18}$$

In the following sections, it is proved that Eq. (5.1) is nonlinearly self-adjoint and conserved vectors associated with different symmetries of this equation are constructed.

5.1.4 Nonlinear self-adjointness of linear TFDE

With no loss of generality, one can set $k = 1$ in (5.1). It was shown in [67] that, for both considered types of fractional derivatives and all $\alpha \in (0, 2)$, the corresponding Lie algebra of point symmetries is infinite and is spanned by generators

$$\begin{aligned} X_1 &= \frac{\partial}{\partial x}, \qquad & X_2 &= 2t\frac{\partial}{\partial t} + \alpha x\frac{\partial}{\partial x}, \\ X_3 &= u\frac{\partial}{\partial u}, \qquad & X_\infty &= h\frac{\partial}{\partial u}, \end{aligned} \tag{5.19}$$

where $h = h(t, x)$ is an arbitrary solution of the equation $\mathcal{D}_t^\alpha h = h_{xx}$.

In the considered linear case, the adjoint Eq. (5.9) takes the form

$$(\mathcal{D}_t^\alpha)^* v = v_{xx}. \tag{5.20}$$

It can be seen that this equation is also linear and does not contain the function u.

Let $v = \varphi(t, x) \neq 0$ be an arbitrary nontrivial solution of this adjoint equation. Because (5.20) is satisfied upon the substitution $v = \varphi(t, x)$ for all $u(t, x)$ then in accordance with the definition of nonlinear self-adjointness given in the previous section, the linear Eq. (5.1) is nonlinearly self-adjoint with such a function $\varphi(t, x)$. The adjoint Eq. (5.20) has nontrivial solutions. For example, particular nontrivial solutions of this equation are $v(t, x) = ct^{\alpha-1}x$ for $\mathcal{D}_t^\alpha = {}_0^{RL}D_t^\alpha$, and $v(t, x) = ctx$ for $\mathcal{D}_t^\alpha = {}_0^C D_t^\alpha$ (here, c is an arbitrary constant).

The formal Lagrangian (5.5) for the linear Eq. (5.1) has the form

$$\mathcal{L} = \varphi(t, x) \left[\mathcal{D}_t^\alpha u - u_{xx}\right]. \tag{5.21}$$

Using this Lagrangian, one can find the conserved vectors for linear Eq. (5.1) corresponding to the symmetries (5.19) by the formulae (5.18) with the Noether operators defined by (5.13)–(5.15).

5.1.5 Conservation laws for TFDE with the Riemann–Liouville fractional derivative

Calculations by (5.18) give the following results for the components of conserved vectors for Eq. (5.1) with the Riemann–Liouville fractional derivative.

For the subdiffusion equation when $\alpha \in (0,1)$ the components of conserved vectors are given by

$$\begin{aligned} C_i^t &= \varphi_0 I_t^{1-\alpha}(\mathcal{W}_i) + \mathfrak{J}(\mathcal{W}_i, \varphi_t), \\ C_i^x &= \varphi_x \mathcal{W}_i - \varphi \mathcal{W}_{ix}. \end{aligned}$$

Here, subscript i coincides with the number of appropriate symmetry from (5.19) ($i = 1, 2, 3$ and ∞), and functions $\mathcal{W}_i$ have the form

$$\begin{aligned} &\mathcal{W}_1 = u_x, \quad \mathcal{W}_2 = 2tu_t + \alpha x u_x, \\ &\mathcal{W}_3 = u, \qquad\qquad \mathcal{W}_\infty = h. \end{aligned} \tag{5.22}$$

In the same way as for the diffusion-wave equation when $\alpha \in (1,2)$, the components of conserved vectors are given by

$$\begin{aligned} C_i^t &= \varphi\, {}_0 I_t^{\alpha-1}(\mathcal{W}_i) - \varphi_t\, {}_0 I_t^{2-\alpha}(\mathcal{W}_i) - \mathfrak{J}(\mathcal{W}_i, \varphi_{tt}), \\ C_i^x &= \varphi_x \mathcal{W}_i - \varphi \mathcal{W}_{ix}, \qquad i = 1, 2, 3, \infty. \end{aligned}$$

Because operators D_x and ${}_0^{RL}D_t^\alpha$ commute with each other then the conserved vectors corresponding to generators X_1 and X_2 can be rewritten in another form. Define

$$w = 2t\varphi_t + \alpha x \varphi_x. \tag{5.23}$$

Case 1 Subdiffusion equation ($0 < \alpha < 1$)

$$\begin{aligned} X_1 : C_1^t &= \varphi_x\, {}_0 I_t^{1-\alpha} u + \mathfrak{J}(u, \varphi_{tx}), \\ C_1^x &= -\varphi_x u_x + \varphi_{xx} u; \\ X_2 : C_2^t &= w\, {}_0 I_t^{1-\alpha} u - 2t \left[\varphi_t\, {}_0 I_t^{1-\alpha} u - u\, {}_t I_T^{1-\alpha} \varphi_t\right] \\ &\quad + 2\mathfrak{J}(tu_t - (\alpha - 1)u, \varphi_t) - \alpha x \mathfrak{J}(u, \varphi_{tx}), \\ C_2^x &= -w u_x + w_x u. \end{aligned}$$

Case 2 Diffusion-wave equation $(1 < \alpha < 2)$

$$
\begin{aligned}
X_1 : \mathcal{C}_1^t &= \varphi_x \, {}_0^{RL}D_t^{\alpha-1} u - \varphi_{tx} \, {}_0I_t^{2-\alpha} u - \mathfrak{J}(u, \varphi_{ttx}), \\
\mathcal{C}_1^x &= -\varphi_x u_x + \varphi_{xx} u; \\
X_2 : \mathcal{C}_2^t &= w \, {}_0^{RL}D_t^{\alpha-1} u - w_t \, {}_0I_t^{2-\alpha} u \\
&\quad + 2t \left[\varphi_{tt} \, {}_0I_t^{2-\alpha} u - u \, {}_tI_T^{2-\alpha} \varphi_{tt} \right] \\
&\quad + 2\mathfrak{J}(tu_t - (\alpha - 1)u, \varphi_{tt}) - \alpha x \mathfrak{J}(u, \varphi_{ttx}), \\
\mathcal{C}_2^x &= -w u_x + w_x u.
\end{aligned}
$$

5.1.6 Conservation laws for TFDE with the Caputo fractional derivative

For the subdiffusion equation when $\alpha \in (0, 1)$, one can find

$$
\begin{aligned}
\mathcal{C}_i^t &= \mathcal{W}_i \, {}_tI_T^{1-\alpha}(\varphi) - \mathfrak{J}(\mathcal{W}_{it}, \varphi), \\
\mathcal{C}_i^x &= \varphi_x \mathcal{W}_i - \varphi \mathcal{W}_{ix}, \qquad i = 1, 2, 3, \infty.
\end{aligned}
$$

For the diffusion-wave equation when $\alpha \in (1, 2)$, the components of conserved vectors can be written as

$$
\begin{aligned}
\mathcal{C}_i^t &= \mathcal{W}_i \, {}_t^{RL}D_T^{\alpha-1} \varphi + \mathcal{W}_{it} \, {}_tI_T^{2-\alpha} \varphi - \mathfrak{J}(\mathcal{W}_{itt}, \varphi), \\
\mathcal{C}_i^x &= \varphi_x \mathcal{W}_i - \varphi \mathcal{W}_{ix}, \qquad i = 1, 2, 3, \infty.
\end{aligned}
$$

As previously, the functions $\mathcal{W}_i$ are defined by (5.22). Similar to the case of the Riemann–Liouville fractional derivative, the conserved vectors corresponding to generators X_1 and X_2 can be presented in another form.
Case 1 Subdiffusion equation $(0 < \alpha < 1)$

$$
\begin{aligned}
X_1 : \mathcal{C}_1^t &= u \, {}_tI_T^{1-\alpha} \varphi_x - \mathfrak{J}(u_t, \varphi_x), \\
\mathcal{C}_1^x &= -\varphi_x u_x + \varphi_{xx} u; \\
X_2 : \mathcal{C}_2^T &= -\frac{2T}{\Gamma(1-\alpha)} \frac{u\varphi(T, x)}{(T-t)^\alpha} \\
&\quad + u \, {}_tI_T^{1-\alpha} w - 2t \left[u_t \, {}_tI_T^{1-\alpha} \varphi - \varphi \, {}_0I_t^{1-\alpha} u_t \right] \\
&\quad + 2\mathfrak{J}(tu_{tt} - (\alpha - 2)u_t, \varphi) - \alpha x \mathfrak{J}(u_t, \varphi_x), \\
\mathcal{C}_2^x &= -w u_x + w_x u.
\end{aligned}
$$

Case 2 Diffusion-wave equation ($1 < \alpha < 2$)

$$\begin{aligned}
X_1 : C_1^t &= u\, {}_t^{RL}D_T^{\alpha-1}\varphi_x + u_t\, {}_tI_T^{2-\alpha}\varphi_x - \mathfrak{J}(u_{tt}, \varphi_x), \\
C_1^x &= -\varphi_x u_x + \varphi_{xx} u; \\
X_2 : C_2^T &= -\frac{2T}{\Gamma(1-\alpha)} \frac{u\varphi(T,x)}{(T-t)^\alpha} \\
&\quad - \frac{2T}{\Gamma(2-\alpha)} \frac{u_t\varphi(T,x)}{(T-t)^\alpha} \\
&\quad + u_t\, {}_tI_T^{2-\alpha} w + u\, {}_t^{RL}D_T^{\alpha-1} w \\
&\quad - 2t\left[u_{tt}\, {}_tI_T^{2-\alpha}\varphi - \varphi\, {}_0I_t^{2-\alpha} u_{tt}\right] \\
&\quad + 2\mathfrak{J}(tu_{ttt} - (\alpha-3)u_{tt}, \varphi) - \alpha x \mathfrak{J}(u_{tt}, \varphi_x), \\
C_2^x &= -wu_x + w_x u.
\end{aligned}$$

Here, w is defined by (5.23).

5.1.7 Symmetries and nonlinear self-adjointness of nonlinear TFDE

Now, let us consider a general case when the diffusion coefficient $k(u) \neq$ const. As it was shown in [67], in both cases of the Riemann–Liouville and Caputo time-fractional derivatives of order $\alpha \in (0,2)$ and for arbitrary $k(u)$, Eq. (5.1) has a two-dimensional Lie algebra of point symmetries spanned by generators

$$X_1 = \frac{\partial}{\partial x}, \quad X_2 = 2t\frac{\partial}{\partial t} + \alpha x \frac{\partial}{\partial x}. \tag{5.24}$$

This algebra extends in some special cases of $k(u)$. If $k(u) = u^\beta$ ($\beta \neq 0$), then Eq. (5.1) has an additional symmetry

$$X_3^{(1)} = \beta x \frac{\partial}{\partial x} + 2u\frac{\partial}{\partial u} \tag{5.25}$$

for any $\alpha \in (0,2)$. If $\beta = -\frac{4}{3}$, i.e., $k(u) = u^{-\frac{4}{3}}$, there is an additional extension

$$X_4^{(1)} = x^2\frac{\partial}{\partial x} - 3xu\frac{\partial}{\partial u}. \tag{5.26}$$

Equation (5.1) with the Riemann–Liouville fractional derivative also admits the generator

$$X_4^{(2)} = t^2\frac{\partial}{\partial t} + (\alpha-1)tu\frac{\partial}{\partial u} \tag{5.27}$$

for $k(u) = u^\beta$ with $\beta = -2\alpha/(\alpha-1), \alpha \in (0,2)$.

Finally, if $k(u) = e^u$, Eq. (5.1) with the Caputo fractional derivative of order $\alpha \in (0, 2)$ has the symmetry

$$X_3^{(2)} = x\frac{\partial}{\partial x} + 2\frac{\partial}{\partial u}. \tag{5.28}$$

Contrary to the case of $k = \text{const}$, for the considered case of $k = k(u)$, the adjoint Eq. (5.9) depends on the function u. Nevertheless, there are specific solutions of this equation that do not depend on this function.

If Eq. (5.1) with the Riemann–Liouville fractional derivative is considered, then the right-sided Caputo fractional derivative is used in the corresponding adjoint Eq. (5.9). The particular solutions of this adjoint equation, which are valid for any solution u of Eq.(5.1), have the form

$$v(t,x) = c_1 + c_2 x \quad \text{for } \alpha \in (0,1), \tag{5.29}$$
$$v(t,x) = c_1 + c_2 x + (c_3 + c_4 x)t \quad \text{for } \alpha \in (1,2). \tag{5.30}$$

Here, $c_i (i = 1, 2, 3, 4)$ are arbitrary constants. If Eq. (5.1) with the Caputo fractional derivative is considered, then the right-sided Riemann–Liouville fractional derivative is used in the corresponding adjoint Eq. (5.9). This adjoint equation has the following particular solutions:

$$v(t,x) = (T-t)^{\alpha-1}(c_1 + c_2 x) \quad \text{for } \alpha \in (0,1), \tag{5.31}$$
$$v(t,x) = (T-t)^{\alpha-2} \times [c_1 + c_3 x + (T-t)(c_2 + c_4 x)] \quad \text{for } \alpha \in (1,2). \tag{5.32}$$

These solutions are also valid for every solution u of Eq. (5.1) with the Caputo time-fractional derivative.

Contrary to the solutions (5.29) and (5.30), the solutions (5.31) and (5.32) depend on the right time boundary T.

So, one can declare that nonlinear time-fractional subdiffusion and diffusion-wave equations with the Riemann–Liouville and Caputo fractional derivatives are nonlinearly self-adjoint. Therefore, the solutions (5.29)–(5.32) can be substituted into the formal Lagrangian (5.5) which then can be used for constructing conservation laws.

5.1.8 Conservation laws for nonlinear TFDE with the Riemann–Liouville fractional derivative

Using the Noether operators (5.13) and (5.15), the symmetries (5.24)–(5.26) and the formal Lagrangian (5.5) with the function $v(t,x)$ given by (5.29), only one new conserved vector has been found for Eq. (5.1) with the Riemann–Liouville time-fractional derivative of order $\alpha \in (0,1)$. This conserved vector has the components

$$C^t = x\, {}_0I_t^{1-\alpha} u, \qquad C^x = K(u) - xk(u)u_x, \tag{5.33}$$

where $K(u)$ is an arbitrary function such that $K'(u) = k(u)$. Note that the operator X_1 produces the trivial conserved vector for the constant c_1 from (5.29), and the conserved vector (5.3) for the constant c_2. The operators X_2 and X_3 give (5.3) for the constant c_1, and (5.33) for the constant c_2. The operator $X_4^{(1)}$ gives (5.33) for the constant c_1 and the trivial conserved vector for the constant c_2.

In the case of $k(u) = u^{\frac{2\alpha}{1-\alpha}}$, using the symmetry (28), two new conservation laws have been found. Corresponding conserved vectors have the components

$$C^t = t\,{}_0I_t^{1-\alpha}u - {}_0I_t^{2-\alpha}u, \qquad C^x = -tu^{\frac{2\alpha}{1-\alpha}}u_x \tag{5.34}$$

and

$$\begin{aligned} C^t &= x\left(t\,{}_0I_t^{1-\alpha}u - {}_0I_t^{2-\alpha}u\right) \\ C^x &= tu^{\frac{2\alpha}{1-\alpha}}\left(\frac{1-\alpha}{1+\alpha}u - xu_x\right). \end{aligned} \tag{5.35}$$

The conserved vector (5.34) corresponds to the constant c_1, and the conserved vector (5.35) corresponds to the constant c_2 from (5.29).

For the nonlinear diffusion-wave equation with the Riemann–Liouville time-fractional derivative, five new conservation laws have been found. The components of corresponding conserved vectors are presented in Table 5.1, where the conserved vector number 1 is the known conserved vector (5.3). As previously, in this table, $K(u)$ is an arbitrary function such that $K'(u) = k(u)$.

The correspondence between the symmetries (5.24)– (5.27), the constants $c_i (i = 1, 2, 3, 4)$ from (5.30) and the conserved vectors numbers from Table 5.1 is established by Table 5.2. In this table, index 0 corresponds to the trivial conserved vectors.

It is interesting to note that contrary to the linear case, the conserved vectors for the nonlinear TFDE (5.1) with the Riemann–Liouville time-fractional derivative do not involve the integral (5.16). Moreover, the obtained conserved vectors for the nonlinear TFDE do not depend on the right time boundary T.

5.1.9 Conservation laws for nonlinear TFDE with the Caputo fractional derivative

Using the Noether operators (5.14) and (5.15), the symmetries (5.24)–(5.26), (5.28), and the formal Lagrangian (5.5) with the function $v(t, x)$ given by (5.31), four new conservation laws have been found for the subdiffusion equation (5.1) with the Caputo fractional derivative. The corresponding

TABLE 5.1: Conserved vectors for the diffusion-wave equation with the Riemann–Liouville fractional derivative

No.	Components of the conserved vectors
1.	$C^t = {}_0^{RL}D_t^{\alpha-1}u$ $C^x = -k(u)u_x$
2.	$C^t = t\,{}_0^{RL}D_t^{\alpha-1}u - {}_0I_t^{2-\alpha}u$ $C^x = -tk(u)u_x$
3.	$C^t = x\,{}_0^{RL}D_t^{\alpha-1}u$ $C^x = K(u) - xk(u)u_x$
4.	$C^t = tx\,{}_0^{RL}D_t^{\alpha-1}u - x\,{}_0I_t^{2-\alpha}u$ $C^x = tK(u) - txk(u)u_x$
5.	$C^t = t^2\,{}_0^{RL}D_t^{\alpha-1}u - 2t\,{}_0I_t^{2-\alpha}u + 2\,{}_0I_t^{3-\alpha}u$ $C^x = -t^2k(u)u_x$
6.	$C^t = t^2x\,{}_0^{RL}D_t^{\alpha-1}u - 2tx\,{}_0I_t^{2-\alpha}u + 2x\,{}_0I_t^{3-\alpha}u$ $C^x = t^2K(u) - t^2xk(u)u_x$

TABLE 5.2: The correspondence between symmetries and conserved vectors numbers for the diffusion-wave equation with the Riemann–Liouville fractional derivative

	X_1	X_2	$X_3^{(1)}$	$X_4^{(1)}$	$X_4^{(2)}$
c_1	0	1	1	3	2
c_2	1	3	3	0	4
c_3	0	2	2	4	5
c_4	2	4	4	0	6

TABLE 5.3: Conserved vectors for the subdiffusion equation with the Caputo fractional derivative

No.	Components of the conserved vectors
1.	$C^t = u(0,x)\Phi(t) + (T-t)^{\alpha}{}_0I_t^{1-\alpha}\left(\frac{u}{T-t}\right)$ $C^x = -(T-t)^{\alpha-1}k(u)u_x$
2.	$C^t = (T-t)^{\alpha-1}{}_0I_t^{2-\alpha}\left(\frac{u_t}{T-t}\right)$ $C^x = -(T-t)^{\alpha-2}k(u)u_x$
3.	$C^t = xu(0,x)\Phi(t) + x(T-t)^{\alpha}{}_0I_t^{1-\alpha}\left(\frac{u}{T-t}\right)$ $C^x = (T-t)^{\alpha-1}[K(u) - xk(u)u_x]$
4.	$C^t = x(T-t)^{\alpha-1}{}_0I_t^{2-\alpha}\left(\frac{u_t}{T-t}\right)$ $C^x = (T-t)^{\alpha-2}[K(u) - xk(u)u_x]$

TABLE 5.4: The correspondence between symmetries and conserved vectors numbers for the subdiffusion equation with the Caputo fractional derivative

	X_1	X_2	$X_3^{(1)}$	$X_3^{(2)}$	$X_4^{(1)}$
c_1	0	1, 2	1	1	3
c_2	1	3, 4	3	3	0

conserved vectors are presented in Table 5.3, where the function $\Phi(t)$ is

$$\Phi(t) = \frac{1}{\alpha\Gamma(1-\alpha)}\left(1-\frac{t}{T}\right)^{\alpha} \times {}_2F_1\left(\alpha,\alpha;\alpha+1;1-\frac{t}{T}\right).$$

Here, ${}_2F_1(\,,\,;\,;\,)$ is the Gauss hypergeometric function. The correspondence between the symmetries (5.24)–(5.26), (5.28), the constants c_1 and c_2 from (5.30), and the conserved vectors numbers from Table 5.3 is established by Table 5.4. Thus, the symmetry X_2 produces all four conservation laws. The trivial conserved vectors are produced by the operator X_1 for the constant c_1 and by the operator $X_4^{(1)}$ for the constant c_2. In Table 5.4, the trivial conserved vectors are denoted by 0.

For nonlinear diffusion-wave equation with the Caputo time-fractional derivative, six new conserved vectors have been found. The components of these vectors are presented in Table 5.5. As previously, in this table

TABLE 5.5: Conserved vectors for the diffusion-wave equation with the Caputo time-fractional derivative

No.	Components of the conserved vectors
1.	$C^t = (T-t)^{\alpha-2} {}_0I_t^{3-\alpha}\left(\frac{u_{tt}}{T-t}\right)$ $C^x = -(T-t)^{\alpha-3}k(u)u_x$
2.	$C^t = \Phi(t)u_t(0,x) + (T-t)^{\alpha-1} {}_0I_t^{2-\alpha}\left(\frac{u_t}{T-t}\right)$ $C^x = -(T-t)^{\alpha-2}k(u)u_x$
3.	$C^t = \Psi(t)u_t(0,x) + (T-t)^{\alpha} {}_0^FI_t^{2-\alpha}\left(\frac{u_t}{T-t}\right)$ $C^x = -(T-t)^{\alpha-1}k(u)u_x$
4.	$C^t = x(T-t)^{\alpha-2} {}_0I_t^{3-\alpha}\left(\frac{u_{tt}}{T-t}\right)$ $C^x = (T-t)^{\alpha-3}(K(u) - xk(u)u_x)$
5.	$C^t = x\Phi(t)u_t(0,x) + x(T-t)^{\alpha-1} {}_0I_t^{2-\alpha}\left(\frac{u_t}{T-t}\right)$ $C^x = (T-t)^{\alpha-2}(K(u) - xk(u)u_x)$
6.	$C^t = x\Psi(t)u_t(0,x) + x(T-t)^{\alpha} {}_0^FI_t^{2-\alpha}\left(\frac{u_t}{T-t}\right)$ $C^x = (T-t)^{\alpha-1}(K(u) - xk(u)u_x)$

$K'(u) = k(u)$. Also, the following notations are used in Table 5.5:

$$\Phi(t) = \frac{1}{(\alpha-1)\Gamma(2-\alpha)}\left(1-\frac{t}{T}\right)^{\alpha-1} \times {}_2F_1\left(\alpha-1, \alpha-1; \alpha; 1-\frac{t}{T}\right),$$

$$\Psi(t) = \frac{1}{\alpha\Gamma(2-\alpha)}\left(1-\frac{t}{T}\right)^{\alpha} \times {}_2F_1\left(\alpha-1, \alpha; \alpha+1; 1-\frac{t}{T}\right),$$

$$\left({}_0^FI_t^{2-\alpha}f\right)(t) = \frac{1}{\Gamma(2-\alpha)}\int_0^t \frac{f(\tau)}{(t-\tau)^{\alpha-1}} \times {}_2F_1\left(1,1;2-\alpha;\frac{t-\tau}{T-\tau}\right)d\tau$$

(here, $1 < \alpha < 2$).

The correspondence between the symmetries (5.24)–(5.26), (5.28), the constants c_i ($i = 1,2,3,4$) from (5.32), and the conserved vectors numbers from Table 5.5 is established by Table 5.6. Thus, the symmetry X_2 produces all six conservation laws. The trivial conserved vectors are produced by the operator X_1 for the constants c_1 and c_2, and by the operator $X_4^{(1)}$ for the constants c_3 and c_4. In Table 5.6, the trivial conserved vectors are denoted by 0.

TABLE 5.6: The correspondence between symmetries and conserved vectors numbers for the diffusion-wave equation with the Caputo fractional derivative

	X_1	X_2	$X_3^{(1)}$	$X_3^{(2)}$	$X_4^{(1)}$
c_1	0	1, 2	2	2	5
c_2	0	2, 3	3	3	6
c_3	2	4, 5	5	5	0
c_4	3	5, 6	6	6	0

Finally, one additional remark should be made. If $u_t(0, x) = 0$, then the operator $X_4^{(2)}$ given by (5.27) is admitted by Eq. (5.1) with the Caputo time-fractional derivative of order $\alpha \in (1, 2)$ and $k(u) = u^{\frac{2\alpha}{1-\alpha}}$. So, this operator can be considered as a conditional symmetry for this equation. This operator produces all six conserved vectors from Table 5.5: vectors 1, 2, 3 for the constant c_1, vectors 2, 3 for the constant c_2, vectors 4, 5, 6 for the constant c_3 and vectors 5, 6 for the constant c_4.

5.2 Conservation laws of fractional diffusion-absorption equation

Time-fractional version of the diffusion-absorption (TFDA) equation has the following form:

$$\partial_t^\alpha u = u u_{xx} + \frac{1}{\sigma}(u_x)^2 - \sigma, \quad \alpha \in (0, 1). \tag{5.36}$$

Similar to the previous section, we use the Ibragimov method [95] for constructing the conservation laws of Eq. (5.36). Formal Lagrangian of the TFDA equation can be written as:

$$\mathcal{L} = v(x, t)\left[\partial_t^\alpha u - u u_{xx} - \frac{1}{\sigma} u_x^2 + \sigma\right], \tag{5.37}$$

where $v(x, t)$ denotes the dependent nonlocal variable. We can construct the adjoint equation to the TFDA equation as Euler-Lagrange equation:

$$\frac{\delta \mathcal{L}}{\delta u} = (D_t^\alpha)^* v - u v_{xx} - \left(\frac{2}{\sigma} - 2\right)(v u_x)_x = 0. \tag{5.38}$$

The components of the conserved vectors for Eq. (5.36) are

$$\mathcal{C}_i^t = vD_t^{\alpha-1}(\mathcal{W}_i) + \mathfrak{J}\left(\mathcal{W}_i, v_t\right), \tag{5.39}$$

$$\mathcal{C}_i^x = \mathcal{W}_i\left(v_x u + (1-\frac{2}{\sigma})vu_x\right) - vuD_x(\mathcal{W}_i), \quad i=1,2,3, \tag{5.40}$$

where

$$\mathcal{W}_1 = -u_x, \quad \mathcal{W}_2 = -\alpha u - tu_t, \quad \mathcal{W}_3 = 2u - xu_x. \tag{5.41}$$

That is

$$\begin{aligned}
\mathcal{C}_1^t &= vD_t^{\alpha-1}(-u_x) + \mathfrak{J}\left(-u_x, v_t\right),\\
\mathcal{C}_1^x &= -u_x\left(v_x u + (1-\frac{2}{\sigma})vu_x\right) + vuu_{xx},\\
\mathcal{C}_2^t &= -\alpha vD_t^{\alpha-1}(u) - tvD_t^{\alpha-1}(u_t) - (\alpha-1)vD_t^{\alpha-2}(u_t) - \alpha\mathfrak{J}\left(u, v_t\right) - \mathfrak{J}\left(tu_t, v_t\right),\\
\mathcal{C}_2^x &= -(\alpha u + tu_t)\left(v_x u + (1-\frac{2}{\sigma})vu_x\right) + vu(\alpha u_x - tu_{tx}),
\end{aligned} \tag{5.42}$$

and

$$\begin{aligned}
\mathcal{C}_3^t &= 2vD_t^{\alpha-1}(u) - xvD_t^{\alpha-1}(u_x) + 2\mathfrak{J}\left(u, v_t\right) - \mathfrak{J}\left(xu_x, v_t\right),\\
\mathcal{C}_3^x &= (2u - xu_x)\left(v_x u + (1-\frac{2}{\sigma})vu_x\right) - vuu_x + xvuu_{xx}.
\end{aligned} \tag{5.43}$$

5.3 Nonlinear self-adjointness of the Kompaneets equations

In the same manner as it was done for the equations of integer orders, we define the formal Lagrangian $\mathcal{L}$ for the Eq. (2.87) by

$$\mathcal{L} = vF(t, x, u, \mathcal{D}_t^{\gamma(\alpha)}u, u_x, u_{xx}). \tag{5.44}$$

The adjoint equation to the Eq. (2.87) is defined by

$$F^*\left(t, x, u, v, \mathcal{D}_t^{\gamma(\alpha)}u, \left(\mathcal{D}_t^{\gamma(\alpha)}\right)^* v, u_x, v_x, u_{xx}, v_{xx}\right) \equiv \frac{\partial \mathcal{L}}{\partial u} = 0. \tag{5.45}$$

Here $\left(\mathcal{D}_t^{\gamma(\alpha)}\right)^*$ denotes the adjoint operator to $D_t^{\gamma(\alpha)}$. It is defined below for each particular case of fractional differential operators used in the Eq. (2.87). If we consider the Eq. (2.87) for a finite time interval $t \in [0, T]$, then the corresponding Euler-Lagrange operator $\frac{\partial}{\partial u}$ in (5.45) has the form

$$\frac{\partial}{\partial u} = \frac{\partial}{\partial u} - D_x\frac{\partial}{\partial u_x} + D_x^2\frac{\partial}{\partial u_{xx}} + \left(\mathcal{D}_t^{\gamma(\alpha)}\right)^*\frac{\partial}{\partial(\mathcal{D}_t^{\alpha}u)}, \tag{5.46}$$

where

$$\left(\mathcal{D}_t^{\gamma(\alpha)}\right)^* \equiv (D_t^\alpha D_t)^* = {}_tD_T^\alpha D_t$$

for the Eq. (2.82),

$$\left(\mathcal{D}_t^{\gamma(\alpha)}\right)^* \equiv \left({}^C D_t^\alpha\right)^* = {}_tD_T^\alpha$$

for the Eq. (2.84),

$$\left(\mathcal{D}_t^{\gamma(\alpha)}\right)^* \equiv \left(D_t^{1+\alpha}\right)^* = {}_t^C D_T^{1+\alpha}$$

for the Eq. (2.85) and

$$\left(\mathcal{D}_t^{\gamma(\alpha)}\right)^* \equiv (D_t^\alpha)^* = {}_t^C D_T^\alpha$$

for the Eq. (2.86). Here

$${}_tD_T^\beta u = \frac{(-1)^n}{\Gamma(n-\beta)} \frac{\partial^n}{\partial t^n} \int_t^T \frac{u(\tau, x)}{(\tau - t)^{\beta - n + 1}} d\tau$$

is the right-sided Riemann-Liouville time-fractional derivative of order $\beta \in R_+$, $n = [\beta] + 1$, and

$${}_t^C D_T^\beta u = \frac{(-1)^n}{\Gamma(n-\beta)} \int_t^T \frac{D_\tau^n u(\tau, x)}{(\tau - t)^{\beta - n + 1}} d\tau$$

is the right-sided Caputo time-fractional derivative of order $\beta \in R_+$, $n = [\beta] + 1$.

After simple calculations in (5.45), we obtain the following adjoint time-fractional Kompaneets equation

$$\left(\mathcal{D}_t^{\gamma(\alpha)}\right)^* v - x^2 v_{xx} + x^2(1 + 2u)v_x + 2(1 - x - 2xu)v = 0 \tag{5.47}$$

for the Eq. (2.87).

The definition of the nonlinear self-adjointness can be extended to the time-fractional Kompaneets equations. Namely, the Eq. (2.87) is said to be nonlinearly self-adjoint if the adjoint Eq. (5.47) is satisfied for all solutions u of the Eq. (2.87) upon a substitution

$$v = \varphi(t, x, u) \tag{5.48}$$

satisfying the condition $\varphi(t, x, u) \neq 0$.

We find all substitutions (5.48) that provide the nonlinear self-adjointness of the time-fractional Kompaneets Eqs. (2.82)-(2.86) and their approximations, and arrive at the following result.

Lemma 5.3.1 *The time-fractional Kompaneets Eq. (2.87) and their diffusion-type approximations are all nonlinearly self-adjoint and the substitution (5.48) has the form*

$$\varphi = \Theta(t)\Psi(x). \tag{5.49}$$

Here the function $\Phi(t)$ depends on the type of fractional differential operator $D_t^{\gamma(\alpha)}$, namely

$$\mathcal{D}_t^{\gamma(\alpha)} = D_t^{\alpha} D_t : \quad \Phi(t) = \phi_1 (T-t)^{\alpha} + \phi_2; \tag{5.50}$$

$$\mathcal{D}_t^{\gamma(\alpha)} =^C D_t^{\alpha} D_t : \quad \Phi(t) = \phi_1 (T-t)^{\alpha-1}; \tag{5.51}$$

$$\mathcal{D}_t^{\gamma(\alpha)} = D_t^{1+\alpha} : \quad \Phi(t) = \phi_1 t + \phi_2; \tag{5.52}$$

$$\mathcal{D}_t^{\gamma(\alpha)} = D_t^{\alpha} : \quad \Phi(t) = \phi_1, \tag{5.53}$$

where ϕ_1 and ϕ_2 are arbitrary constants. The function $\Psi(x)$ depends on the approximation of the function $h(u, u_x)$ defined by (2.88), namely

$$h = u_x + u + u^2, h = u_x + u^2 : \Psi(x) = \psi_1 x^2; \tag{5.54}$$

$$h = u_x : \Psi(x) = \psi_1 x^2 + \psi_2 x^{-1}; \tag{5.55}$$

$$h = u_x + u : \Psi(x) = \psi_1 x^2 + \psi_2 \Theta(x), \tag{5.56}$$

where ψ_1 and w2 are arbitrary constants, and the function $\Theta(x)$ is

$$\Theta(x) = e^x x^{-1} \left[e^{-x} x^3 Ei(x) - x^2 - x - 2 \right]. \tag{5.57}$$

Remark 5.3.1 *We do not present here the derivation of the substitutions (5.49). Instead, we note that the term $\psi_1 x^2$ in each substitution is precisely the similar substitution for the classical Kompaneets Eq. (2.77). The functions $\Theta(t)$ defined by (5.50)-(5.53) are the solutions of the fractional equation $D_t^{\gamma(\alpha)}\varphi = 0$. The functions $\Psi(x)$ defined by (5.54)-(5.56) are the solutions of the equation*

$$x^2\Psi'' - x^2 h_u \Psi' + 2(x h_u - 1)\Psi = 0$$

for arbitrary function u.

Now, we extend the usual notion of a conserved vector to the time-fractional Eq. (2.87). A vector $\mathcal{C} = (\mathcal{C}^t, \mathcal{C}^x)$ is called a conserved vector for the Eq. (2.87) if it satisfies the conservation equation

$$\left[D_t(\mathcal{C}^t) + D_x(\mathcal{C}^x) \right]_{(2.9)} = 0.$$

Let the Eq. (2.87) be nonlinearly self-adjoint and admit a one-parameter point transformation group with the generator

$$X = \xi^1(t,x,u)\frac{\partial}{\partial t} + \xi^2(t,x,u)\frac{\partial}{\partial x} + \phi(t,x,u)\frac{\partial}{\partial u}.$$

Since the Eq. (2.87) does not involve the fractional derivatives with respect to x, the x-component of the conserved vector can be found by the general formula for calculating conserved vectors associated with symmetries. This formula gives

$$\mathcal{C}^x = \mathcal{W}\left(\frac{\partial \mathcal{L}}{\partial u_x} - D_x \frac{\partial \mathcal{L}}{\partial u_{xx}}\right) + D_x(\mathcal{W})\frac{\partial \mathcal{L}}{\partial u_{xx}}. \tag{5.58}$$

Here

$$\mathcal{W} = \phi - \xi^1 u_t - \xi^2 u_x,$$

and $\mathcal{L}$ is the formal Lagrangian (5.44) where the variable v is eliminated by using a suitable substitution $v = \varphi(t, x, u)$.
The formula for the t-component of conserved vector depends on the type of time-fractional derivative in the Eq. (2.87):

$$\mathcal{C}^t = I_t^{1-\alpha} D_t(\mathcal{W})\frac{\partial \mathcal{L}}{\partial (D_t^\alpha u_t)} - \mathcal{W} I_T^{1-\alpha} D_t \frac{\partial \mathcal{L}}{\partial (D_t^\alpha u_t)} + \mathfrak{J}\left(D_t(\mathcal{W}), D_t \frac{\partial \mathcal{L}}{\partial (D_t^\alpha u_t)}\right) \tag{5.59}$$

for the Eq. (2.82),

$$\mathcal{C}^t = \mathcal{W} I_T^{1-\alpha}\frac{\partial \mathcal{L}}{\partial ({}^C D_t^\alpha u)} - \mathfrak{J}\left(D_t(\mathcal{W}), \frac{\partial \mathcal{L}}{\partial ({}^C D_t^\alpha u)}\right) \tag{5.60}$$

for the Eq. (2.84),

$$\mathcal{C}^t = D_t^\alpha(\mathcal{W})\frac{\partial \mathcal{L}}{\partial (D_t^{1+\alpha} u)} - I_t^{1-\alpha}(\mathcal{W}) D_t \frac{\partial \mathcal{L}}{\partial (D_t^{1+\alpha} u)} - \mathfrak{J}\left(\mathcal{W}, D_t^2 \frac{\partial \mathcal{L}}{\partial (D_t^{1+\alpha} u)}\right) \tag{5.61}$$

for the Eq. (2.85) and

$$\mathcal{C}^t = I_t^{1-\alpha}(\mathcal{W})\frac{\partial \mathcal{L}}{\partial (D_t^\alpha u)} + \mathfrak{J}\left(\mathcal{W}, D_t \frac{\partial \mathcal{L}}{\partial (D_t^\alpha u)}\right) \tag{5.62}$$

for the Eq. (2.86).
Now we consider separately each of the Eqs. (2.82)-(2.86).

5.3.1 Conservation laws for approximations of the Eq. (2.82)

In this case we substitute $D_t^\gamma(\alpha) = D_t^\alpha D_t$ in the Eq. (2.87). Then this equation has the conservation form with the conserved vector having the components

$$\mathcal{C}^t = x^2 I_t^{1-\alpha} u_t, \qquad \mathcal{C}^x = -x^4 h(u, u_x). \tag{5.63}$$

The formal Lagrangian (5.44) upon the substitution $v = \varphi(t, x)$, with the function $\varphi(t, x)$ defined by (5.49),(5.50), takes the form

$$\mathcal{L} = (\phi_1 (T-t)^\alpha + \phi_2)\Psi(x)(D_t^\alpha u_t - x^2 D_x(h) - 4xh). \tag{5.64}$$

Substituting the formal Lagrangian (5.64) and the symmetries (2.89)-(2.91) in (5.58) and (5.59) we find new conservation laws.
For $h = u_x$, $h = u_x + u$, $h = u_x + u^2$ we obtain the conserved vector

$$\begin{aligned} \mathcal{C}^t &= x^2\left[(T-t)^\alpha I_t^{1-\alpha}u_t + \Gamma(1+\alpha)u - \alpha\mathfrak{J}(u_t,(T-t)^{\alpha-1})\right], \\ \mathcal{C}^x &= -x^4(T-t)^\alpha h(u,u_x). \end{aligned} \tag{5.65}$$

The conserved vector (5.65) corresponds to the product $\phi_1\psi_1$ of the constants ϕ_1 and ψ_1 in the substitution (5.49) with (5.50),(5.54). The conserved vector corresponding to the product $\phi_2\psi_1$ of the constants ϕ_2 and ψ_1 in (5.49) coincides with the conserved vector (5.63).
For $h = u_x$, using the symmetry X_1 from (2.89), we find two additional conserved vectors:

$$\begin{aligned} \mathcal{C}^t &= x^{-1}\left[(T-t)^\alpha I_t^{1-\alpha}u_t + \Gamma(1+\alpha)u - \alpha I(u_t,(T-t)^{\alpha-1})\right], \\ \mathcal{C}^x &= -(T-t)^\alpha(xu_x + 3u), \end{aligned} \tag{5.66}$$

$$\mathcal{C}^t = x^{-1}J_t^{1-\alpha}u_t, \qquad \mathcal{C}^x = -(xu_x + 3u). \tag{5.67}$$

The conserved vector (5.66) corresponds to the product $\phi_1\psi_2$ of the constants ϕ_1 and ψ_2 in (5.49) with (5.50), (5.55), and conserved vector (5.67) corresponds to the product $\phi_2\psi_2$ of the constants ϕ_2 and ψ_2. Note that the symmetry X_2 from (2.89) provides the trivial conservation law only.
For $h = u_x + u$, using the symmetry X_1 from (2.90), we also find two additional conserved vectors:

$$\begin{aligned} \mathcal{C}^t &= \Theta(x)\left[(T-t)^\alpha I_t^{1-\alpha}u_t + \Gamma(1+\alpha)u - \alpha\mathfrak{J}(u_t,(T-t)^{\alpha-1})\right], \\ \mathcal{C}^x &= -(T-t)^\alpha x^2\left[\Theta(x)u_x + (\Theta(x) + 2x^{-1}\Theta(x) - \Theta(x))u\right], \end{aligned} \tag{5.68}$$

$$\begin{aligned} \mathcal{C}^t &= \Theta(x)I_t^{1-\alpha}u_t, \\ \mathcal{C}^x &= -x^2\left[\Theta(x)u_x + (\Theta(x) + 2x^{-1}\Theta(x) - \Theta;(x))u\right], \end{aligned} \tag{5.69}$$

where the function $\Theta(x)$ is defined by (5.57). The conserved vector (5.68) corresponds to the product $\phi_1\psi_2$ of the constants ϕ_1 and ψ_2 in (5.49) with (5.50), (5.56), and conserved vector (5.69) corresponds to the product $\phi_2\psi_2$ of the constants ϕ_2 and ψ_2.

5.3.2 Conservation laws for approximations of the Eq. (2.84)

In this case we substitute $D_t^{\gamma(\alpha)} = {}^C D_t^\alpha$ in the Eq. (2.87). Unlike the previous case, the equation under consideration does not have a conservation form. The formal Lagrangian (5.44) upon the substitution $v = \varphi(t,x)$, with the function $\varphi(t,x)$ defined by (5.49), (5.51), takes the form

$$\mathcal{L} = \phi_1(T-t)^{\alpha-1}\Psi(x)({}^C D_t^\alpha u - x^2 D_x(h) - 4xh). \tag{5.70}$$

Substituting the formal Lagrangian (5.70) and the symmetries (2.89)-(2.91) in (5.58) and (5.60) we find new conservation laws.
For $h = u_x$, $h = u_x + u$, $h = u_x + u^2$ we obtain the conserved vector

$$\begin{aligned} \mathcal{C}^t &= x^2 \left[\Gamma(\alpha)u - \mathfrak{J}(u_t, (T-t)^{\alpha-1})\right], \\ \mathcal{C}^x &= -x^4 (T-t)^{\alpha-1} h(u, u_x). \end{aligned} \tag{5.71}$$

The conserved vector (5.71) corresponds to the product $\phi_1\psi_1$ of the constants ϕ_1 and ψ_1 in the substitution (5.49) with (5.51), (5.54).
For $h = u_x$, using the symmetry X_1 from (2.89), we find the conserved vector

$$\begin{aligned} \mathcal{C}^t &= x^{-1} \left[\Gamma(\alpha)u - \mathfrak{J}(u_t, (T-t)^{\alpha-1})\right], \\ \mathcal{C}^x &= -(T-t)^{\alpha}(xu_x + 3u). \end{aligned} \tag{5.72}$$

The conserved vector (5.72) corresponds to the product $\phi_1\psi_2$ of the constants ϕ_1 and ψ_2 in (5.49) with (5.51), (5.55). The symmetry X_2 from (2.89) provides the trivial conservation law only.
For $h = u_x + u$, using the symmetry X_1 from (2.90), we also find an additional conserved vector, namely:

$$\begin{aligned} \mathcal{C}^t &= \Theta(x) \left[\Gamma(\alpha)u - \mathfrak{J}(u_t, (T-t)^{\alpha-1})\right], \\ \mathcal{C}^x &= -(T-t)^{\alpha} x^2 \left[\Theta(x)u_x + (\Theta(x) + 2x^{-1}\Theta(x) - \Theta'(x))u\right], \end{aligned} \tag{5.73}$$

where the function $\Theta(x)$ is defined by (5.57). The conserved vector (5.73) corresponds to the product $\phi_1\psi_2$ of the constants ϕ_1 and ψ_2 in (5.49) with (5.51), (5.56).

5.3.3 Conservation laws for approximations of the Eq. (2.85)

Now we consider the Eq. (2.85) upon substitution $D_t^{\gamma(\alpha)}$. Then this equation has the conservation form with the conserved vector having the components

$$\mathcal{C}^t = x^2 D_t^{\alpha} u, \qquad \mathcal{C}^x = -x^4 h(u, u_x). \tag{5.74}$$

The formal Lagrangian (5.44) upon the substitution $v = \varphi(t, x)$, with the function $\varphi(t, x)$ defined by (5.49), (5.52), takes the form

$$\mathcal{L} = (\phi_1 t + \phi_2)\Psi(x)(D_t^{1+\alpha}u - x^2 D_x(h) - 4xh). \tag{5.75}$$

Substituting the formal Lagrangian (5.75) and the symmetries (2.89)-(2.91) in (5.58) and (5.61), we find new conserved vectors for all three particular types of the function $h(u, u_x)$.
For $h = u_x$, $h = u_x + u$, $h = u_x + u^2$ we obtain the conserved vector

$$\begin{aligned} \mathcal{C}^t &= x^2 \left[tD_t^{\alpha}u - I_t^{1-\alpha}u\right], \\ \mathcal{C}^x &= -tx^4 h(u, u_x). \end{aligned} \tag{5.76}$$

The conserved vector (5.76) corresponds to the product $\phi_1\psi_1$ of the constants ϕ_1 and ψ_1 in (5.49) with (5.52), (5.54). The conserved vector corresponds to the product $\phi_2\psi_1$ of the constants ϕ_2 and ψ_1 in (5.49) coinciding with the conserved vector (5.64).
For $h = u_x$, using the symmetry X_1, we find two additional conserved vectors:

$$\begin{aligned} \mathcal{C}^t &= x^{-1}\left[tD_t^\alpha u - I_t^{1-\alpha}u\right], \\ \mathcal{C}^x &= -t(xu_x + 3u), \end{aligned} \tag{5.77}$$

$$\mathcal{C}^t = x^{-1}D_t^\alpha u, \qquad \mathcal{C}^x = -(xu_x + 3u). \tag{5.78}$$

The conserved vector (5.77) corresponds to the product $\phi_1\psi_2$ of the constants ϕ_1 and ψ_2 in (5.49) with (5.52), (5.55), and conserved vector (5.78) corresponds to the product $\phi_2\psi_2$ of the constants ϕ_2 and ψ_2. As previously, the symmetry X_2 from (2.89) provides the trivial conservation law only.
For $h = u_x + u$, using the symmetry X_1 from (2.90), we also construct two additional conserved vectors:

$$\begin{aligned} \mathcal{C}^t &= \Theta(x)\left[tD_t^\alpha u - I_t^{1-\alpha}u\right], \\ \mathcal{C}^x &= -tx^2\left[\Theta(x)u_x + (\Theta(x) + 2x^{-1}\Theta(x) - \Theta'(x))u\right]; \end{aligned} \tag{5.79}$$

$$\begin{aligned} \mathcal{C}^t &= \Theta(x)D_t^\alpha u, \\ \mathcal{C}^x &= -x^2\left[\Theta(x)u_x + (\Theta(x) + 2x^{-1}\Theta(x) - \Theta'(x))u\right], \end{aligned} \tag{5.80}$$

where the function $\Theta(x)$ is defined by (5.57). The conserved vector (5.79) corresponds to the product $\phi_1\psi_2$ of the constants ϕ_1 and ψ_2 in (5.49) with (5.52), (5.56), and conserved vector (5.80) corresponds to the product $\phi_2\psi_2$ of the constants ϕ_2 and ψ_2.

5.3.4 Conservation laws for approximations of the Eq. (2.86)

Now we substitute $\mathcal{D}_t^\alpha = D_t^\alpha$ in the Eq. (2.87). Then this equation has the conservation form with the conserved vector having the components

$$\mathcal{C}^t = x^2 I_t^{1-\alpha}u, \qquad \mathcal{C}^x = -x^4 h(u, u_x). \tag{5.81}$$

The formal Lagrangian (5.44) upon the substitution $v = \varphi(t,x)$, with the function $\varphi(t,x)$ defined by (5.49), (5.53), takes the form

$$\mathcal{L} = \phi_1\Psi(x)(D_t^\alpha u - x^2 D_x(h) - 4xh). \tag{5.82}$$

Substituting the formal Lagrangian (5.82) and the symmetries (2.89)-(2.91) in (5.58) and (5.62) we construct two new conservation laws.
For $h = u_x$, using the symmetry X_1 from (2.89), we find a conserved vector

$$\mathcal{C}^t = x^{-1}I_t^{1-\alpha}u, \qquad \mathcal{C}^x = -(xu_x + 3u). \tag{5.83}$$

The conserved vector (5.83) corresponds to the product $\phi_1\psi_2$ of the constants ϕ_1 and ψ_2 in (5.49) with (5.53), (5.55). The symmetry X_2 from (2.89) provides the trivial conservation law only.
For $h = u_x + u$, using the symmetry X_1 from (2.90), we also construct two additional conserved vectors:

$$\begin{aligned} C^t &= \Theta(x) I_t^{1-\alpha} u, \\ C^x &= -x^2 \left[\Theta(x) u_x + (\Theta(x) + 2x^{-1}\Theta(x) - \Theta'(x))u\right], \end{aligned} \tag{5.84}$$

where the function $\Theta(x)$ is defined by (5.57). The conserved vector (5.84) corresponds to the product $\phi_1\psi_2$ of the constants ϕ_1 and ψ_2 in (5.49) with (5.53), (5.56).
It is interesting to note that for all three particular types of the function $h(u, u_x)$, i.e., $h = u_x$, $h = u_x + u^2$, the symmetries (2.89)-(2.91) give the conserved vector with the components (5.81) that correspond to the product $\phi_1\psi_1$ of the constants ϕ_1 and ψ_1 in the substitution (5.49) with (5.53), (5.54).

5.3.5 Noninvariant particular solutions

One can use the method of conservation laws (see [92], chapter 2) for constructing the particular solutions for the diffusion-type approximations of the time-fractional Kompaneets equation. According to this method, a particular solution is obtained by letting

$$C^t = p(x), \quad C^x = q(t),$$

where C^t and C^x are the components of a conserved vector. The calculations lead to the following results.
Calculations show that solutions of any approximation of the Eq. (2.87), obtained by using conservation laws presented in the previous section and corresponding this approximation, coincide.
For all approximations of the time-fractional Kompaneets Eqs. (2.87) with $h = u_x$, this approach gives the linear combinations of the invariant solutions corresponding to $\beta = 0$ and $\beta = -3$ that have been presented in the previous subsection. For $h = u_x$ and $h = u_x + u^2$, using the method of conservation laws, we obtain the following solutions:

- for the approximation of the Eq. (2.82)

$$u = z(x, c_1, c_2)t^\alpha + z(x, c_3, c_4),$$

- for the approximation of the Eq. (2.84)

$$u = z(x, c_1, c_2),$$

- for the approximation of the Eq. (2.85)

$$u = (x, c_1, c_2)t^{\alpha-1} + z(x, c_3, c_4)t^\alpha,$$

- for the approximation of the Eq. (2.86)

$$u = z(x, c_1, c_2)t^{\alpha-1},$$

where c_1; c_2; c_3; c_4 are arbitrary constants. The function z depends on the function h : for $h = u_x + u$ we have

$$z(x, a, b) = e^{-x}\left[ax^{-2}\Theta(x) + b\right],$$

where the function $\Theta(x)$ is defined by (5.57). For $h = u_x + u^2$ we find two types of the function f:

$$z(x, a, b) = \frac{1}{x} + \frac{a}{x^2}\tanh\left(b - \frac{a}{x}\right), \quad z(x, a, b) = \frac{a}{x} - \frac{a}{x^2}\tan\left(b - \frac{a}{x}\right).$$

5.4 Conservation laws of the time-fractional CRW equation

Like the above section, the formal Lagrangian of the time-fractional CRW equation (2.72) can be investigated as:

$$\mathcal{L} = v(x,t)\left[\partial_t^\alpha u + u_{xx} - u_x + 2uu_x\right], \tag{5.85}$$

where $v(x, t)$ denotes a dependent nonlocal variable. Moreover, we can construct the adjoint equation as the Euler-Lagrange equation:

$$\frac{\delta\mathcal{L}}{\delta u} = (D_t^\alpha)^* v + (1 - 2u)v_x + v_{xx} = 0. \tag{5.86}$$

Therefore, the components of the conserved vectors for Eq. (2.72) are

$$\mathcal{C}_i^t = vD_t^{\alpha-1}(\mathcal{W}_i) + \mathfrak{J}\left(\mathcal{W}_i, v_t\right), \tag{5.87}$$

$$\mathcal{C}_i^x = \mathcal{W}_i\left(v(1 - 2u) - v_x\right) + vD_x(\mathcal{W}_i), \quad i = 1, 2, \tag{5.88}$$

where

$$\mathcal{W}_1 = -u_x, \quad \mathcal{W}_2 = \alpha(1 - 2u) - 4tu_t - 2\alpha xu_x. \tag{5.89}$$

That is

$$\begin{aligned}
\mathcal{C}_1^t &= vD_t^{\alpha-1}(-u_x) + \mathfrak{J}\left(-u_x, v_t\right),\\
\mathcal{C}_1^x &= -u_x\left(v(1 - 2u) - v_x\right) - vu_{xx},\\
\mathcal{C}_2^t &= -2\alpha D_t^{\alpha-1}(u) - 4vD_t^{\alpha-1}(tu_t) - 2\alpha vD_t^{\alpha-1}(xu_x)\\
&\quad +\alpha\mathfrak{J}\left(1 - 2u, v_t\right) - 4\mathfrak{J}\left(tu_t, v_t\right) - 2\alpha\mathfrak{J}\left(xu_x, v_t\right),\\
\mathcal{C}_2^x &= \left(\alpha(1 - 2u) - 4tu_t - 2\alpha xu_x\right)\left(v(1 - 2u) - v_x\right)\\
&\quad -v\left(4\alpha u_x + 4tu_{tx} + 2\alpha xu_{xx}\right).
\end{aligned} \tag{5.90}$$

5.5 Conservation laws of the time-fractional VB equation and time-fractional BB equation

In this section, after some preliminaries we obtain the conservation laws [95, 80] for the equations

$$\begin{cases} \partial_t^\alpha u + v_x + uu_x = 0, \\ \partial_t^\alpha v + (uv)_x + u_{xxx} = 0, \end{cases} \tag{5.91}$$

and

$$\begin{cases} \partial_t^\alpha u - \frac{1}{2}v_x + 2uu_x = 0, \\ \partial_t^\alpha v - \frac{1}{2}u_{xxx} + 2(uv)_x = 0, \end{cases} \tag{5.92}$$

respectively.

Consider a time-fractional k^{th}-order PDE of two independent variables $x,\ t$ and dependent variables $u,\ v$, viz.:

$$\begin{cases} \mathcal{F}_1(x,t,u,v,\partial_t^\alpha u,\partial_t^\alpha v,u_x,\dots,u_{kx},v_x,\dots,v_{kx}) = 0, \\ \mathcal{F}_2(x,t,u,v,\partial_t^\alpha u,\partial_t^\alpha v,u_x,\dots,u_{kx},v_x,\dots,v_{kx}) = 0 \end{cases}, \quad \alpha \in (0,1). \tag{5.93}$$

The formal Lagrangian of system (5.93) can be written as

$$\mathcal{L} = \psi(x,t)\mathcal{F}_1 + \theta(x,t)\mathcal{F}_2, \tag{5.94}$$

where $\psi(x,t) = \Psi(x,t,u,v)$ and $\theta(x,t) = \Theta(x,t,u,v)$ are new dependent variables. Corresponding system of adjoint equations is defined as follows:

$$\begin{cases} \mathcal{F}_1^* \equiv \frac{\delta\mathcal{L}}{\delta u} = 0, \\ \mathcal{F}_2^* \equiv \frac{\delta\mathcal{L}}{\delta v} = 0, \end{cases} \tag{5.95}$$

where the Euler-Lagrange operators are given by the formal sums:

$$\frac{\delta}{\delta u} = \frac{\partial}{\partial u} + (\partial_t^\alpha)^* \frac{\partial}{\partial(\partial_t^\alpha u)} + \sum_{s\geq 1}(-1)^s \mathcal{D}_x \cdots \mathcal{D}_x \frac{\partial}{\partial u_{sx}}, \tag{5.96}$$

and

$$\frac{\delta}{\delta v} = \frac{\partial}{\partial v} + (\partial_t^\alpha)^* \frac{\partial}{\partial(\partial_t^\alpha v)} + \sum_{s\geq 1}(-1)^s \mathcal{D}_x \cdots \mathcal{D}_x \frac{\partial}{\partial v_{sx}}, \tag{5.97}$$

with

$$\mathcal{D}_x = \frac{\partial}{\partial x} + u_x \frac{\partial}{\partial u} + v_x \frac{\partial}{\partial v} + u_{xx} \frac{\partial}{\partial u_x} + v_{xx} \frac{\partial}{\partial v_x} + \cdots \tag{5.98}$$

being the total derivative operators with respect to x.
Moreover, $(\partial_t^\alpha)^*$ is the adjoint operator defined by

$$(\partial_t^\alpha)^* = -I_C^{1-\alpha}\left(\frac{d}{dt}\right) = {}_t^C D_C^\alpha, \tag{5.99}$$

with $I_C^{1-\alpha}$ being the right fractional integral operator and ${}_t^C D_C^{\alpha}$ being the right Caputo fractional derivative.
The time-fractional system (5.93) is called nonlinearly self-adjoint if [80]

$$\begin{cases} \frac{\delta \mathcal{L}}{\delta u} = \mu_1 \mathcal{F}_1 + \mu_2 \mathcal{F}_2, \\ \frac{\delta \mathcal{L}}{\delta v} = \mu_3 \mathcal{F}_1 + \mu_4 \mathcal{F}_2, \end{cases} \tag{5.100}$$

where $\mu_i,\ i = 1, \ldots, 4$ are unknowns to be determined.
Also, if $\mathfrak{A}$ is the set of all differential functions of all finite orders, and $\xi,\ {}^u\phi,\ {}^v\phi \in \mathfrak{A}$, then *Lie-Bäcklund* operator is

$$X = \tau \frac{\partial}{\partial t} + \xi \frac{\partial}{\partial x} + {}^u\phi \frac{\partial}{\partial u} + {}^v\phi \frac{\partial}{\partial v} + {}^u\zeta \frac{\partial}{\partial u_x} + {}^v\zeta \frac{\partial}{\partial v_x} + \cdots, \tag{5.101}$$

where

$$\begin{aligned} {}^u\zeta &= \mathcal{D}_x\big({}^u\phi\big) - u_x \mathcal{D}_x(\xi) - u_t \mathcal{D}_t \tau, \\ {}^v\zeta &= \mathcal{D}_x\big({}^v\phi\big) - v_x \mathcal{D}_x(\xi) - v_t \mathcal{D}_t \tau. \end{aligned}$$

One can write the *Lie-Bäcklund* operator (5.101) in characteristic form

$$\begin{aligned} X = \tau \mathcal{D}_t + \xi \mathcal{D}_x + {}^u\mathcal{W} \frac{\partial}{\partial u} + {}^v\mathcal{W} \frac{\partial}{\partial v} + \sum_{s \geq 1} \mathcal{D}_x \ldots \mathcal{D}_x({}^u\mathcal{W}) \frac{\partial}{\partial u_{sx}} \\ + \sum_{s \geq 1} \mathcal{D}_x \ldots \mathcal{D}_x({}^v\mathcal{W}) \frac{\partial}{\partial v_{sx}}, \end{aligned} \tag{5.102}$$

where ${}^u\mathcal{W} = {}^u\phi - \xi u_x - \tau u_t$ and ${}^v\mathcal{W} = {}^v\phi - \xi v_x - \tau v_t$ are the characteristic functions.
The two-tuple vector $T = (T^t, T^x),\ T^t, T^x \in \mathfrak{A}$ is a conserved vector of Eq. (5.93) if

$$\mathcal{D}_t(T^t) + \mathcal{D}_x(T^x) = 0, \tag{5.103}$$

on the solution space of (5.93).

Theorem 25 *Every Lie point, Lie-Bäcklund and nonlocal symmetry admitted by the Eq. (5.93) gives rise to a conservation law for the system consisting of the Eq. (5.93) and the adjoint Eq. (5.95) where the components T^t and T^x of the conserved vector $T = (T^t, T^x)$ are determined by*

$$\begin{aligned} T^t = D_t^{\alpha-1}({}^u\mathcal{W}) \frac{\partial \mathcal{L}}{\partial(\partial_t^\alpha u)} + D_t^{\alpha-1}({}^v\mathcal{W}) \frac{\partial \mathcal{L}}{\partial(\partial_t^\alpha v)} \\ + \mathfrak{I}\left({}^u\mathcal{W}, D_t \frac{\partial \mathcal{L}}{\partial(\partial_t^\alpha u)}\right) + \mathfrak{I}\left({}^v\mathcal{W}, D_t \frac{\partial \mathcal{L}}{\partial(\partial_t^\alpha v)}\right), \end{aligned} \tag{5.104}$$

and

$$\begin{aligned} T^x = {}^u\mathcal{W} \frac{\delta \mathcal{L}}{\delta u_x} + {}^v\mathcal{W} \frac{\delta \mathcal{L}}{\delta v_x} + \sum_{s \geq 1} \mathcal{D}_x \ldots \mathcal{D}_x({}^u\mathcal{W}) \frac{\delta \mathcal{L}}{\delta u_{(s+1)x}} \\ + \sum_{s \geq 1} \mathcal{D}_x \ldots \mathcal{D}_x({}^v\mathcal{W}) \frac{\delta \mathcal{L}}{\delta v_{(s+1)x}}, \end{aligned} \tag{5.105}$$

where

$$\Im(f,g) = \frac{1}{\Gamma(1-\alpha)} \int_0^t \int_t^T \frac{f(\tau,x)g(\mu,x)}{(\mu-\tau)^\alpha} d\mu d\tau. \tag{5.106}$$

5.5.1 Construction of conservation laws for Eq. (5.91)

The formal Lagrangian for the system of (5.91) can be written as

$$\mathcal{L} = \psi(x,t)\left(\partial_t^\alpha u + v_x + uu_x\right) + \theta(x,t)\left(\partial_t^\alpha v + (uv)_x + u_{xxx}\right), \tag{5.107}$$

which admits the adjoint equation

$$\begin{cases} \mathcal{F}_1^* = (\partial_t^\alpha)^*\psi - \psi_x u - \theta_x v - \theta_{xxx}, \\ \mathcal{F}_2^* = (\partial_t^\alpha)^*\theta - \psi_x - \theta_x u, \end{cases} \tag{5.108}$$

where

$$\begin{aligned} &\psi_x = \Psi_x + \Psi_u u_x + \Psi_v v_x, \\ &\theta_x = \Theta_x + \Theta_u u_x + \Theta_v v_x, \\ &\theta_{xxx} = \Theta_{xxx} + 3\Theta_{uuv}u_x^2 v_x + 3\Theta_{uvv}u_x v_x^2 + 3\Theta_{uu}u_x u_{xx} + 3\Theta_{uv}u_x v_{xx} \\ &+ 3\Theta_{uv}u_{xx}v_x + 3\Theta_{vv}v_x v_{xx} + 6\Theta_{xuv}u_x v_x + 3\Theta_{xxu}u_x + 3\Theta_{xxv}v_x + 3\Theta_{xu}u_{xx} \\ &+ 3\Theta_{xv}v_{xx} + \Theta_u u_{xxx} + \Theta_v v_{xxx} + 3\Theta_{xuu}u_x^2 + 3\Theta_{xvv}v_x^2 + \Theta_{uuu}u_x^3 + \Theta_{vvv}v_x^3. \end{aligned}$$

Therefore, the nonlinear self-adjoint condition (5.100) can be written as

$$\begin{aligned} &(\partial_t^\alpha)^*\psi - \left(\Psi_x + \Psi_u u_x + \Psi_v v_x\right)u - \left(\Theta_x + \Theta_u u_x + \Theta_v v_x\right)v - \Theta_{xxx} \\ &- 3\Theta_{uuv}u_x^2 v_x - 3\Theta_{uvv}u_x v_x^2 - 3\Theta_{uu}u_x u_{xx} - 3\Theta_{uv}u_x v_{xx} - 3\Theta_{uv}u_{xx}v_x \\ &- 3\Theta_{vv}v_x v_{xx} - 6\Theta_{xuv}u_x v_x - 3\Theta_{xxu}u_x - 3\Theta_{xxv}v_x - 3\Theta_{xu}u_{xx} \\ &- 3\Theta_{xv}v_{xx} - \Theta_u u_{xxx} - \Theta_v v_{xxx} - 3\Theta_{xuu}u_x^2 - 3\Theta_{xvv}v_x^2 - \Theta_{uuu}u_x^3 \\ &- \Theta_{vvv}v_x^3 = \mu_1\left(\partial_t^\alpha u + v_x + uu_x\right) + \mu_2\left(\partial_t^\alpha v + (uv)_x + u_{xxx}\right), \end{aligned} \tag{5.109}$$

and

$$\begin{aligned} &(\partial_t^\alpha)^*\theta - \left(\Psi_x + \Psi_u u_x + \Psi_v v_x\right) - \left(\Theta_x + \Theta_u u_x + \Theta_v v_x\right)u \\ &= \mu_3\left(\partial_t^\alpha u + v_x + uu_x\right) + \mu_4\left(\partial_t^\alpha v + (uv)_x + u_{xxx}\right). \end{aligned} \tag{5.110}$$

Solving Eqs. (5.109)-(5.110) yields

$$\begin{aligned} &\mu_i = 0, \quad i = 1,\ldots,4, \\ &\Psi(x,t,u,v) = A, \quad \Theta(x,t,u,v) = B, \quad A, B \in \mathbb{R}. \end{aligned}$$

Therefore, if we suppose $A = B = 1$, then

$$\mathcal{L} = \partial_t^\alpha u + \partial_t^\alpha v + v_x + uu_x + (uv)_x + u_{xxx}. \tag{5.111}$$

We recall that Eq. (5.91) admits a two-dimensional Lie algebras; thus we consider the following two cases:
(i) We first consider the Lie point symmetry generator $X_1 = \frac{\partial}{\partial x}$. Corresponding characteristic functions are

$$^u\mathcal{W} = -u_x, \quad ^v\mathcal{W} = -v_x. \tag{5.112}$$

By using the Theorem 25, the components of the conserved vector are given by

$$\begin{aligned} T^t =& - I_t^{\alpha-1}(u_x + v_x), \\ T^x =& - u_x(u+v) - v_x(1+u) - u_{xxx}. \end{aligned}$$

(ii) Using Lie point symmetry generator $X_2 = 2t\frac{\partial}{\partial t} + \alpha x\frac{\partial}{\partial x} - \alpha u\frac{\partial}{\partial u} - 2\alpha v\frac{\partial}{\partial v}$ and Theorem 25, one can obtain the conserved vector whose components are

$$\begin{aligned} T^t =& - \alpha I_t^{1-\alpha}(u+2v) - \alpha x I_t^{1-\alpha}(u_x + v_x) - 2I_t^{1-\alpha}(tu_t + tv_t), \\ T^x =& - (\alpha u + \alpha x u_x + 2tu_t)(u+v) - (2\alpha v + \alpha x v_x + 2tv_t)(1+u) \\ & - \alpha(3u_{xx} + xu_{xxx}) - 2tu_{txx}. \end{aligned}$$

In this case, the corresponding characteristic functions are

$$^u\mathcal{W} = -\alpha u - \alpha x u_x - 2tu_t, \quad ^v\mathcal{W} = -2\alpha v - \alpha x v_x - 2tv_t. \tag{5.113}$$

Bibliography

[1] S. Abbas, M. Benchohra, and G.M. N'Guérékata. *Topics in fractional differential equations*, volume 27. Springer Science & Business Media, 2012.

[2] S. Abbasbandy and M.S. Hashemi. Group preserving scheme for the Cauchy problem of the Laplace equation. *Engineering Analysis with Boundary Elements*, 35(8):1003–1009, 2011.

[3] S. Abbasbandy, M.S. Hashemi, and I. Hashim. On convergence of homotopy analysis method and its application to fractional integro-differential equations. *Quaestiones Mathematicae*, 36(1):93–105, 2013.

[4] S. Abbasbandy, M.S. Hashemi, and C.-S. Liu. The Lie-group shooting method for solving the Bratu equation. *Communications in Nonlinear Science and Numerical Simulation*, 16(11):4238–4249, 2011.

[5] F. Abidi and K. Omrani. The homotopy analysis method for solving the Fornberg–Whitham equation and comparison with Adomian's decomposition method. *Computers & Mathematics with Applications*, 59(8):2743–2750, 2010.

[6] S.O. Adesanya, M. Eslami, M. Mirzazadeh, and A. Biswas. Shock wave development in couple stress fluid-filled thin elastic tubes. *The European Physical Journal Plus*, 130(6):114, 2015.

[7] O.P. Agrawal. Formulation of Euler–Lagrange equations for fractional variational problems. *Journal of Mathematical Analysis and Applications*, 272(1):368–379, 2002.

[8] O.P. Agrawal. Generalized variational problems and Euler–Lagrange equations. *Computers & Mathematics with Applications*, 59(5):1852–1864, 2010.

[9] A. Akgül and M.S. Hashemi. Group preserving scheme and reproducing kernel method for the Poisson–Boltzmann equation for semiconductor devices. *Nonlinear Dynamics*, 88(4):2817–2829, 2017.

[10] A. Akgül, M.S. Hashemi, and S.A. Raheem. Constructing two powerful methods to solve the Thomas–Fermi equation. *Nonlinear Dynamics*, 87(2):1435–1444, 2017.

[11] S.S. Akhiev and T. Özer. Symmetry groups of the equations with nonlocal structure and an application for the collisionless Boltzmann equation. *International Journal of Engineering Science*, 43(1-2):121–137, 2005.

[12] I. Aliyu, Y. Li, and D. Baleanu. Invariant subspace and classification of soliton solutions of the coupled nonlinear fokas-liu system. *Frontiers in Physics*, 7:39, 2019.

[13] F. Allassia and M.C. Nucci. Symmetries and heir equations for the laminar boundary layer model. *Journal of Mathematical Analysis and Applications*, 201(3):911–942, 1996.

[14] G. AN and J. Cole. The general similarity solution of the heat equation. *Journal of Mathematics and Mechanics*, 18(11):1025–1042, 1969.

[15] D.G. Aronson and H.F. Weinberger. Nonlinear diffusion in population genetics, combustion, and nerve pulse propagation. In *Partial differential equations and related topics*, pages 5–49. Springer, 1975.

[16] T.M. Atanacković, S. Konjik, S. Pilipović, and S. Simić. Variational problems with fractional derivatives: invariance conditions and Noether's theorem. *Nonlinear Analysis: Theory, Methods & Applications*, 71(5-6):1504–1517, 2009.

[17] E.D. Avdonina, N.H. Ibragimov, and R. Khamitova. Exact solutions of gasdynamic equations obtained by the method of conservation laws. *Communications in Nonlinear Science and Numerical Simulation*, 18(9):2359–2366, 2013.

[18] F. Bahrami, R. Najafi, and M.S. Hashemi. On the invariant solutions of space/time-fractional diffusion equations. *Indian Journal of Physics*, 91(12):1571–1579, 2017.

[19] D. Baleanu. About fractional quantization and fractional variational principles. *Communications in Nonlinear Science and Numerical Simulation*, 14(6):2520–2523, 2009.

[20] D. Baleanu, K. Diethelm, and E. Scalas. *Fractional calculus: models and numerical methods*, volume 3. World Scientific, 2012.

[21] D. Baleanu, M. Inc, A. Yusuf, and Aliyu I. Aliyu. Lie symmetry analysis, exact solutions and conservation laws for the time fractional Caudrey–Dodd–Gibbon–Sawada–Kotera equation. *Communications in Nonlinear Science and Numerical Simulation*, 59:222–234, 2018.

[22] D. Baleanu, M. Inc, A. Yusuf, and Aliyu I. Aliyu. Time fractional third-order evolution equation: symmetry analysis, explicit solutions, and conservation laws. *Journal of Computational and Nonlinear Dynamics*, 13(2), 2018.

[23] D. Baleanu, S.I. Muslih, and E.M. Rabei. On fractional Euler–Lagrange and Hamilton equations and the fractional generalization of total time derivative. *Nonlinear Dynamics*, 53(1-2):67–74, 2008.

[24] D. Baleanu, S.I. Muslih, and K. Taş. Fractional Hamiltonian analysis of higher order derivatives systems. *Journal of Mathematical Physics*, 47(10):103503, 2006.

[25] H. Berestycki, O. Diekmann, C.J. Nagelkerke, and P.A. Zegeling. Can a species keep pace with a shifting climate? *Bulletin of Mathematical Biology*, 71(2):399, 2009.

[26] L. Bianchi. *Lezioni sulla teoria dei gruppi continui finiti di trasformazioni.* Enrico Spoerri, 1903.

[27] L. Bianchi. On the three-dimensional spaces which admit a continuous group of motions. *General Relativity and Gravitation*, 33(12):2171–2253, 2001.

[28] G.W. Bluman and S. Kumei. Symmetries and differential equations. Springer. New York, 1989.

[29] G.W. Bluman and S. Kumei. *Symmetries and differential equations*, volume 81. Springer Science & Business Media, 2013.

[30] A.V. Bobylev, G.L. Caraffini, and G. Spiga. On group invariant solutions of the Boltzmann equation. *Journal of Mathematical Physics*, 37(6):2787–2795, 1996.

[31] L. Bourdin, J. Cresson, and I. Greff. A continuous/discrete fractional Noether's theorem. *Communications in Nonlinear Science and Numerical Simulation*, 18(4):878–887, 2013.

[32] B.H. Bradshaw-Hajek, M.P. Edwards, P. Broadbridge, and G.H. Williams. Nonclassical symmetry solutions for reaction–diffusion equations with explicit spatial dependence. *Nonlinear Analysis: Theory, Methods & Applications*, 67(9):2541–2552, 2007.

[33] M.D. Bramson. Maximal displacement of branching Brownian motion. *Communications on Pure and Applied Mathematics*, 31(5):531–581, 1978.

[34] J. Canosa. Diffusion in nonlinear multiplicative media. *Journal of Mathematical Physics*, 10(10):1862–1868, 1969.

[35] T. Cerquetelli, N. Ciccoli, and M.C. Nucci. Four dimensional Lie symmetry algebras and fourth order ordinary differential equations. *Journal of Nonlinear Mathematical Physics*, 9(sup2):24–35, 2002.

[36] S. Chen, F. Liu, P. Zhuang, and V. Anh. Finite difference approximations for the fractional Fokker–Planck equation. *Applied Mathematical Modelling*, 33(1):256–273, 2009.

[37] R. Cherniha. New q-conditional symmetries and exact solutions of some reaction–diffusion–convection equations arising in mathematical biology. *Journal of Mathematical Analysis and Applications*, 326(2):783–799, 2007.

[38] R. Cherniha, M. Serov, and O. Pliukhin. *Nonlinear Reaction-Diffusion-Convection Equations: Lie and Conditional Symmetry, Exact Solutions and Their Applications.* Chapman and Hall/CRC, 2017.

[39] Roman Cherniha, Mykola Serov, and Inna Rassokha. Lie symmetries and form-preserving transformations of reaction–diffusion–convection equations. *Journal of Mathematical Analysis and Applications*, 342(2):1363–1379, 2008.

[40] S. Choudhary, P. Prakash, and V. Daftardar-Gejji. Invariant subspaces and exact solutions for a system of fractional PDEs in higher dimensions. *Computational and Applied Mathematics*, 38(3):126, 2019.

[41] P. A Clarksonz and E.L. Mansfield. Symmetry reductions and exact solutions of a class of nonlinear heat equations. *Physica D: Nonlinear Phenomena*, 70(3):250–288, 1994.

[42] M. Craddock and K.A. Lennox. Lie symmetry methods for multi-dimensional parabolic PDEs and diffusions. *Journal of Differential Equations*, 252(1):56–90, 2012.

[43] L. Debnath. *Nonlinear partial differential equations for scientists and engineers.* Springer Science & Business Media, 2011.

[44] W. Deng. Numerical algorithm for the time fractional Fokker–Planck equation. *Journal of Computational Physics*, 227(2):1510–1522, 2007.

[45] W. Deng. Finite element method for the space and time fractional Fokker–Planck equation. *SIAM Journal on Numerical Analysis*, 47(1):204–226, 2008.

[46] Rosa Di Salvo, M. Gorgone, and F. Oliveri. A consistent approach to approximate Lie symmetries of differential equations. *Nonlinear Dynamics*, 91(1):371–386, 2018.

[47] K. Diethelm. *The analysis of fractional differential equations: An application-oriented exposition using differential operators of Caputo type.* Springer Science & Business Media, 2010.

[48] V.G. Dulov, S.P. Novikov, L.V. Ovsyannikov, B. L. Rozhdestvenskii, A.A. Samarskii, and Y.I. Shokin. Nikolai Nikolaevich Yanenko (obituary). *Russian Mathematical Surveys*, 39(4):99, 1984.

[49] A. Esen. A lumped Galerkin method for the numerical solution of the modified equal-width wave equation using quadratic b-splines. *International Journal of Computer Mathematics*, 83(5-6):449–459, 2006.

[50] A. Esen and S. Kutluay. Solitary wave solutions of the modified equal width wave equation. *Communications in Nonlinear Science and Numerical Simulation*, 13(8):1538–1546, 2008.

[51] R.A. Fisher. The wave of advance of advantageous genes. *Annals of Eugenics*, 7(4):355–369, 1937.

[52] A.S. Fokas. Symmetries and integrability. *Studies in Applied Mathematics*, 77(3):253–299, 1987.

[53] B. Fornberg and G.B. Whitham. A numerical and theoretical study of certain nonlinear wave phenomena. *Philosophical Transactions of the Royal Society of London. Series A, Mathematical and Physical Sciences*, 289(1361):373–404, 1978.

[54] T.D. Frank. Stochastic feedback, nonlinear families of Markov processes, and nonlinear Fokker–Planck equations. *Physica A: Statistical Mechanics and its Applications*, 331(3-4):391–408, 2004.

[55] T.D. Frank. *Nonlinear Fokker-Planck equations: fundamentals and applications.* Springer Science & Business Media, 2005.

[56] G.S.F. Frederico and D.F.M. Torres. A formulation of Noether's theorem for fractional problems of the calculus of variations. *Journal of Mathematical Analysis and Applications*, 334(2):834–846, 2007.

[57] G.S.F. Frederico and D.F.M. Torres. Fractional conservation laws in optimal control theory. *Nonlinear Dynamics*, 53(3):215–222, 2008.

[58] W.I. Fushchich and W.M. Shtelen. On approximate symmetry and approximate solutions of the nonlinear wave equation with a small parameter. *Journal of Physics A: Mathematical and General*, 22(18):L887, 1989.

[59] W.I. Fushchich, W.M. Shtelen, and N.I. Serov. *Symmetry analysis and exact solutions of equations of nonlinear mathematical physics*, volume 246. Springer Science & Business Media, 2013.

[60] V.A. Galaktionov and S.R. Svirshchevskii. *Exact solutions and invariant subspaces of nonlinear partial differential equations in mechanics and physics.* Chapman and Hall/CRC, 2006.

[61] M.L. Gandarias. Weak self-adjoint differential equations. *Journal of Physics A: Mathematical and Theoretical*, 44(26):262001, 2011.

[62] H. Gao, T. Xu, S. Yang, and G. Wang. Analytical study of solitons for the variant Boussinesq equations. *Nonlinear Dynamics*, 88(2):1139–1146, 2017.

[63] R.K. Gazizov and N.H. Ibragimov. Lie symmetry analysis of differential equations in finance. *Nonlinear Dynamics*, 17(4):387–407, 1998.

[64] R.K. Gazizov, N.H. Ibragimov, and S.Y. Lukashchuk. Nonlinear self-adjointness, conservation laws and exact solutions of time-fractional Kompaneets equations. *Communications in Nonlinear Science and Numerical Simulation*, 23(1-3):153–163, 2015.

[65] R.K. Gazizov and A.A. Kasatkin. Construction of exact solutions for fractional order differential equations by the invariant subspace method. *Computers & Mathematics with Applications*, 66(5):576–584, 2013.

[66] R.K. Gazizov, A.A. Kasatkin, and S.Y. Lukashchuk. Continuous transformation groups of fractional differential equations. *Vestnik Usatu*, 9(3):21, 2007.

[67] R.K. Gazizov, A.A. Kasatkin, and S.Y. Lukashchuk. Symmetry properties of fractional diffusion equations. *Physica Scripta*, 2009(T136):014016, 2009.

[68] J. Goard. Generalised conditional symmetries and Nucci's method of iterating the nonclassical symmetries method. *Applied Mathematics Letters*, 16(4):481–486, 2003.

[69] G.A. Guthrie. Constructing Miura transformations using symmetry groups. 1993.

[70] M.S. Hashemi. Constructing a new geometric numerical integration method to the nonlinear heat transfer equations. *Communications in Nonlinear Science and Numerical Simulation*, 22(1-3):990–1001, 2015.

[71] M.S. Hashemi. Group analysis and exact solutions of the time fractional Fokker–Planck equation. *Physica A: Statistical Mechanics and its Applications*, 417:141–149, 2015.

[72] M.S. Hashemi. On Black-Scholes equation; method of heir-equations, nonlinear self-adjointness and conservation laws. *Bulletin of the Iranian Mathematical Society*, 42(4):903–921, 2016.

[73] M.S. Hashemi. A novel simple algorithm for solving the magneto-hemodynamic flow in a semi-porous channel. *European Journal of Mechanics-B/Fluids*, 65:359–367, 2017.

[74] M.S. Hashemi and S. Abbasbandy. A geometric approach for solving Troesch's problem. *Bulletin of the Malaysian Mathematical Sciences Society*, 40(1):97–116, 2017.

[75] M.S. Hashemi and D. Baleanu. Lie symmetry analysis and exact solutions of the time fractional gas dynamics equation. *J. Optoelectron. Adv. Mater.*, 18(3–4):383–388, 2016.

[76] M.S. Hashemi and D. Baleanu. Numerical approximation of higher-order time-fractional telegraph equation by using a combination of a geometric approach and method of line. *Journal of Computational Physics*, 316:10–20, 2016.

[77] M.S. Hashemi and D. Baleanu. On the time fractional generalized Fisher equation: group similarities and analytical solutions. *Communications in Theoretical Physics*, 65(1):11, 2016.

[78] M.S. Hashemi, D. Baleanu, and M. Parto-Haghighi. A Lie group approach to solve the fractional Poisson equation. *Rom. Journ. Phys.*, 60(9–10):1289–1297, 2015.

[79] M.S. Hashemi, D. Baleanu, M. Parto-Haghighi, and E. Darvishi. Solving the time-fractional diffusion equation using a Lie group integrator. *Thermal Science*, 19:S77–S83, 2015.

[80] M.S. Hashemi and Z. Balmeh. On invariant analysis and conservation laws of the time fractional variant Boussinesq and coupled Boussinesq-Burger's equations. *The European Physical Journal Plus*, 133(10):427, 2018.

[81] M.S. Hashemi, Z. Balmeh, and D. Baleanu. Exact solutions, Lie symmetry analysis and conservation laws of the time fractional diffusion-absorption equation. In *Mathematical Methods in Engineering*, pages 97–109. Springer, 2019.

[82] M.S. Hashemi, E. Darvishi, and D. Baleanu. A geometric approach for solving the density-dependent diffusion Nagumo equation. *Advances in Difference Equations*, 2016(1):89, 2016.

[83] M.S. Hashemi, A. Haji-Badali, F. Alizadeh, and D. Baleanu. Integrability, invariant and soliton solutions of generalized Kadomtsev-Petviashvili-modified equal width equation. *Optik*, 139:20–30, 2017.

[84] M.S. Hashemi, A. Haji-Badali, and P. Vafadar. Group invariant solutions and conservation laws of the Fornberg–Whitham equation. *Zeitschrift für Naturforschung A*, 69(8-9):489–496, 2014.

[85] M.S. Hashemi, M. Inc, and A. Akgül. Analytical treatment of the couple stress fluid-filled thin elastic tubes. *Optik*, 145:336–345, 2017.

[86] M.S. Hashemi, M. Inc, E. Karatas, and A. Akgül. A numerical investigation on Burgers equation by MOL-GPS method. *Journal of Advanced Physics*, 6(3):413–417, 2017.

[87] M.S. Hashemi, M. Inc, B. Kilic, and A. Akgül. On solitons and invariant solutions of the magneto-electro-elastic circular rod. *Waves in Random and Complex Media*, 26(3):259–271, 2016.

[88] M.S. Hashemi and M.C. Nucci. Nonclassical symmetries for a class of reaction-diffusion equations: the method of heir-equations. *Journal of Nonlinear Mathematical Physics*, 20(1):44–60, 2013.

[89] M.S. Hashemi, M.C. Nucci, and S. Abbasbandy. Group analysis of the modified generalized Vakhnenko equation. *Communications in Nonlinear Science and Numerical Simulation*, 18(4):867–877, 2013.

[90] M.A.E. Herzallah and D. Baleanu. Fractional-order Euler–Lagrange equations and formulation of Hamiltonian equations. *Nonlinear Dynamics*, 58(1-2):385, 2009.

[91] Peter E. Hydon. *Symmetry methods for differential equations: a beginner's guide*, volume 22. Cambridge University Press, 2000.

[92] N.H. Ibragimov and E.D. Avdonina. Nonlinear self-adjointness, conservation laws, and the construction of solutions of partial differential equations using conservation laws. *Russian Mathematical Surveys*, 68(5):889, 2013.

[93] N.H. Ibragimov and Vladimir F. Kovalev. *Approximate and renormgroup symmetries.* Springer Science & Business Media, 2009.

[94] N.H. Ibragimov. *Elementary Lie group analysis and ordinary differential equations.* Wiley.

[95] N.H. Ibragimov. A new conservation theorem. *Journal of Mathematical Analysis and Applications*, 333(1):311–328, 2007.

[96] N.H. Ibragimov. Quasi-self-adjoint differential equations. *Archives of ALGA*, 4:55–60, 2007.

[97] N.H. Ibragimov. Time-dependent exact solutions of the nonlinear Kompaneets equation. *Journal of Physics A: Mathematical and Theoretical*, 43(50):502001, 2010.

[98] N.H. Ibragimov. Nonlinear self-adjointness and conservation laws. *Journal of Physics A: Mathematical and Theoretical*, 44(43):432002, 2011.

[99] N.H. Ibragimov, V.F. Kovalev, and V.V. Pustovalov. Symmetries of integro-differential equations: A survey of methods illustrated by the Benny equations. *Nonlinear Dynamics*, 28(2):135–153, 2002.

[100] N.Kh Ibragimov. Transformation groups in mathematical physics [in Russian]. 1983.

[101] M. Inc, A. Yusuf, I. Aliyu, and D. Baleanu. Time-fractional Cahn–Allen and time-fractional Klein–Gordon equations: Lie symmetry analysis, explicit solutions and convergence analysis. *Physica A: Statistical Mechanics and its Applications*, 493:94–106, 2018.

[102] B.B. Kadomtsev and O.P. Pogutse. Nonlinear helical perturbations of a plasma in the tokamak. *Zh. Eksp. Teor. Fiz*, 65(5):575–589, 1973.

[103] A.S. Kalashnikov. Some problems of the qualitative theory of nonlinear degenerate second-order parabolic equations. *Russian Mathematical Surveys*, 42(2):169, 1987.

[104] F. Kamenetskii and D. Albertovich. *Diffusion and heat exchange in chemical kinetics*, volume 2171. Princeton University Press, 2015.

[105] K. Khan, M. Ali Akbar, M.M. Rashidi, and I. Zamanpour. Exact traveling wave solutions of an autonomous system via the enhanced (g'/g)-expansion method. *Waves in Random and Complex Media*, 25(4):644–655, 2015.

[106] A. A Kilbas, M. Saigo, and R.K. Saxena. Solution of Volterra integro-differential equations with generalized Mittag-Leffler function in the kernels. *The Journal of Integral Equations and Applications*, pages 377–396, 2002.

[107] A.A. Kilbas, O.I. Marichev, and S.G. Samko. *Fractional integral and derivatives (theory and applications)*. Gordon and Breach, Switzerland, 1993.

[108] A.A. Kilbas, O.I. Marichev, and S.G. Samko. Fractional integral and derivatives (theory and applications), 1993.

[109] A.A. Kilbas, H.M. Srivastava, and J.J. Trujillo. *Theory and applications of fractional differential equations*. North-Holland, New York, 2006.

[110] A.A. Kilbas, H.M. Srivastava, and J.J. Trujillo. *Theory and applications of fractional differential equations*, volume 204. Elsevier Science Limited, 2006.

[111] J. Klafter, S.C. Lim, and R. Metzler. *Fractional dynamics: recent advances*. World Scientific, 2012.

[112] R. Klages, G. Radons, and I.M. Sokolov. *Anomalous transport: foundations and applications*. John Wiley & Sons, 2008.

[113] Hüseyin Koçak, Turgut Öziş, and Ahmet Yıldırım. Homotopy perturbation method for the nonlinear dispersive $K(m,n,1)$ equations with fractional time derivatives. *International Journal of Numerical Methods for Heat & Fluid Flow*, 20(2):174–185, 2010.

[114] A.S. Kompaneets. The establishment of thermal equilibrium between quanta and electrons. *Zhurnal Eksperimentalnoi i Teoreticheskoi Fiziki*, 31:876, 1956.

[115] B. Kruglikov. Symmetry approaches for reductions of PDEs, differential constraints and Lagrange-Charpit method. *Acta Applicandae Mathematicae*, 101(1-3):145–161, 2008.

[116] S. Kumar, A. Kumar, and D. Baleanu. Two analytical methods for time-fractional nonlinear coupled Boussinesq–Burger's equations arise in propagation of shallow water waves. *Nonlinear Dynamics*, 85(2):699–715, 2016.

[117] M. Lakshmanan and P. Kaliappan. Lie transformations, nonlinear evolution equations, and Painlevé forms. *Journal of Mathematical Physics*, 24(4):795–806, 1983.

[118] M.J. Lazo and D.F.M. Torres. The Dubois–Reymond fundamental lemma of the fractional calculus of variations and an Euler–Lagrange equation involving only derivatives of Caputo. *Journal of Optimization Theory and Applications*, 156(1):56–67, 2013.

[119] S. Lie. *Vorlesungen über Differentialgleichungen mit bekannten infinitesimalen Transformationen*. BG Teubner, 1891.

[120] H. Liu and J. Li. Lie symmetry analysis and exact solutions for the short pulse equation. *Nonlinear Analysis: Theory, Methods & Applications*, 71(5-6):2126–2133, 2009.

[121] Y. Liu and D.-S. Wang. Symmetry analysis of the option pricing model with dividend yield from financial markets. *Applied Mathematics Letters*, 24(4):481–486, 2011.

[122] W.-X. Ma and Y. Liu. Invariant subspaces and exact solutions of a class of dispersive evolution equations. *Communications in Nonlinear Science and Numerical Simulation*, 17(10):3795–3801, 2012.

[123] F.M. Mahomed and P.G.L. Leach. Symmetry Lie algebras of nth order ordinary differential equations. *Journal of Mathematical Analysis and Applications*, 151(1):80–107, 1990.

[124] F. Mainardi. *Fractional calculus and waves in linear viscoelasticity: an introduction to mathematical models*. World Scientific, 2010.

[125] A. Majlesi, H.Roohani Ghehsareh, and A. Zaghian. On the fractional Jaulent-Miodek equation associated with energy-dependent Schrödinger potential: Lie symmetry reductions, explicit exact solutions and conservation laws. *The European Physical Journal Plus*, 132(12):516, 2017.

[126] W. Malfliet. Solitary wave solutions of nonlinear wave equations. *American Journal of Physics*, 60(7):650–654, 1992.

[127] A.B. Malinowska. A formulation of the fractional Noether-type theorem for multidimensional Lagrangians. *Applied Mathematics Letters*, 25(11):1941–1946, 2012.

[128] M. Marcelli and M.C. Nucci. Lie point symmetries and first integrals: the Kowalevski top. *Journal of Mathematical Physics*, 44(5):2111–2132, 2003.

[129] S. Martini, N. Ciccoli, and M.C. Nucci. Group analysis and heir-equations of a mathematical model for thin liquid films. *Journal of Nonlinear Mathematical Physics*, 16(01):77–92, 2009.

[130] S.V. Meleshko. Group properties of equations of motions of a viscoelastic medium. *Model. Mekh*, 2(19):114–126, 1988.

[131] S.V. Meleshko. *Methods for constructing exact solutions of partial differential equations: mathematical and analytical techniques with applications to engineering.* Springer Science & Business Media, 2006.

[132] R. Metzler and J. Klafter. The random walk's guide to anomalous diffusion: a fractional dynamics approach. *Physics Reports*, 339(1):1–77, 2000.

[133] A.V. Mikhailov, A.B. Shabat, and V.V. Sokolov. The symmetry approach to classification of integrable equations. In *What is integrability?*, pages 115–184. Springer, 1991.

[134] S. Momani and R. Qaralleh. An efficient method for solving systems of fractional integro-differential equations. *Computers & Mathematics with Applications*, 52(3-4):459–470, 2006.

[135] A.J. Morrison and E.J. Parkes. The n-soliton solution of the modified generalised Vakhnenko equation (a new nonlinear evolution equation). *Chaos, Solitons & Fractals*, 16(1):13–26, 2003.

[136] M.M. Mousa and A. Kaltayev. Application of He's homotopy perturbation method for solving fractional Fokker-Planck equations. *Zeitschrift für Naturforschung A*, 64(12):788–794, 2009.

[137] B. Muatjetjeja and C.M. Khalique. First integrals for a generalized coupled Lane–Emden system. *Nonlinear Analysis: Real World Applications*, 12(2):1202–1212, 2011.

[138] G.M. Mubarakzyanov. On solvable Lie algebras. *Izvestiya Vysshikh Uchebnykh Zavedenii. Matematika*, (1):114–123, 1963.

[139] R. Najafi, F. Bahrami, and M.S. Hashemi. Classical and nonclassical Lie symmetry analysis to a class of nonlinear time-fractional differential equations. *Nonlinear Dynamics*, 87(3):1785–1796, 2017.

[140] M.C. Nucci. Nonclassical symmetries and Bäcklund transformations. *Journal of Mathematical Analysis and Applications*, 178(1):294–300, 1993.

[141] M.C. Nucci. Iterating the nonclassical symmetries method. *Physica D: Nonlinear Phenomena*, 78(1-2):124–134, 1994.

[142] M.C. Nucci. The complete Kepler group can be derived by Lie group analysis. *Journal of Mathematical Physics*, 37(4):1772–1775, 1996.

[143] M.C. Nucci. Iterations of the non-classical symmetries method and conditional Lie-Bäcklund symmetries. *Journal of Physics A: Mathematical and General*, 29(24):8117, 1996.

[144] M.C. Nucci. Nonclassical symmetries as special solutions of heir-equations. *Journal of Mathematical Analysis and Applications*, 279(1):168–179, 2003.

[145] M.C. Nucci. Lie symmetries of a Painlevé-type equation without Lie symmetries. *Journal of Nonlinear Mathematical Physics*, 15(2):205–211, 2008.

[146] M.C. Nucci and P.G.L. Leach. The determination of nonlocal symmetries by the technique of reduction of order. *Journal of Mathematical Analysis and Applications*, 251(2):871–884, 2000.

[147] Z. Odibat and S. Momani. Numerical solution of Fokker–Planck equation with space- and time-fractional derivatives. *Physics Letters A*, 369(5-6):349–358, 2007.

[148] Z.M. Odibat. Solitary solutions for the nonlinear dispersive $K(m, n)$ equations with fractional time derivatives. *Physics Letters A*, 370(3-4):295–301, 2007.

[149] T. Odzijewicz, A. Malinowska, and D. Torres. Noether's theorem for fractional variational problems of variable order. *Open Physics*, 11(6):691–701, 2013.

[150] P.J. Olver. *Applications of Lie groups to differential equations*, volume 107. Springer Science & Business Media, 2012.

[151] T. Özer. Symmetry group classification for two-dimensional elastodynamics problems in nonlocal elasticity. *International Journal of Engineering Science*, 41(18):2193–2211, 2003.

[152] T. Özer. An application of symmetry groups to nonlocal continuum mechanics. *Computers & Mathematics with Applications*, 55(9):1923–1942, 2008.

[153] Georgios Papamikos and Tristan Pryer. A Lie symmetry analysis and explicit solutions of the two-dimensional ∞-polylaplacian. *Studies in Applied Mathematics*, 142(1):48–64, 2019.

[154] S. Pashayi, M.S. Hashemi, and S. Shahmorad. Analytical Lie group approach for solving fractional integro-differential equations. *Communications in Nonlinear Science and Numerical Simulation*, 51:66–77, 2017.

[155] J. Patera and P. Winternitz. Subalgebras of real three- and four-dimensional Lie algebras. *Journal of Mathematical Physics*, 18(7):1449–1455, 1977.

[156] I. Podlubny. *Fractional Differential Equations.* Academic Press, New York, 1999.

[157] R.O. Popovych. Reduction operators of linear second-order parabolic equations. *Journal of Physics A: Mathematical and Theoretical*, 41(18):185202, 2008.

[158] Roman O. Popovych and Nataliya M. Ivanova. New results on group classification of nonlinear diffusion–convection equations. *Journal of Physics A: Mathematical and General*, 37(30):7547, 2004.

[159] Gregory J. Reid. Finding abstract Lie symmetry algebras of differential equations without integrating determining equations. *European Journal of Applied Mathematics*, 2(4):319–340, 1991.

[160] S.Z. Rida, A.M.A. El-Sayed, and A.A.M. Arafa. On the solutions of time-fractional reaction–diffusion equations. *Communications in Nonlinear Science and Numerical Simulation*, 15(12):3847–3854, 2010.

[161] F. Riewe. Nonconservative Lagrangian and Hamiltonian mechanics. *Physical Review E*, 53(2):1890, 1996.

[162] D. Roberts. The general Lie group and similarity solutions for the one-dimensional Vlasov–Maxwell equations. *Journal of Plasma Physics*, 33(2):219–236, 1985.

[163] W. Rudin. *Principles of mathematical analysis.* McGraw-Hill New York, 1964.

[164] H. Rudolf. *Applications of fractional calculus in physics.* World Scientific, 2000.

[165] R.L. Sachs. On the integrable variant of the Boussinesq system: Painlevé property, rational solutions, a related many-body system, and equivalence with the AKNS hierarchy. *Physica D: Nonlinear Phenomena*, 30(1-2):1–27, 1988.

[166] R. Sahadevan and T. Bakkyaraj. Invariant subspace method and exact solutions of certain nonlinear time fractional partial differential equations. *Fractional Calculus and Applied Analysis*, 18(1):146–162, 2015.

[167] A.A. Samarskii, A.P. Mikhailov, et al. *Blow-up in quasilinear parabolic equations*, volume 19. Walter de Gruyter, 2011.

[168] S. Shahmorad, S. Pashaei, and M.S. Hashemi. Numerical solution of a nonlinear fractional integro-differential equation by a geometric approach. *Differential Equations and Dynamical Systems*, pages 1–12, 2017.

[169] Zhang Shi-Hua, Chen Ben-Yong, and Fu Jing-Li. Hamilton formalism and Noether symmetry for mechanico—electrical systems with fractional derivatives. *Chinese Physics B*, 21(10):100202, 2012.

[170] A.F. Sidorov, V.P. Shapeev, and N.N. Ianenko. The method of differential relations and its applications in gas dynamics. *Novosibirsk Izdatel Nauka*, 1984.

[171] V.K. Stokes. Couple stresses in fluids. *The Physics of Fluids*, 9(9):1709–1715, 1966.

[172] N. Taghizadeh, M. Mirzazadeh, and F. Farahrooz. Exact soliton solutions of the modified Kdv–KP equation and the Burgers–KP equation by using the first integral method. *Applied Mathematical Modelling*, 35(8):3991–3997, 2011.

[173] L. Tian and Y. Gao. The global attractor of the viscous Fornberg–Whitham equation. *Nonlinear Analysis: Theory, Methods & Applications*, 71(11):5176–5186, 2009.

[174] H.C. Tuckwell. *Introduction to theoretical neurobiology: volume 2, nonlinear and stochastic theories*, volume 8. Cambridge University Press, 1988.

[175] Y. Uğurlu and D. Kaya. Analytic method for solitary solutions of some partial differential equations. *Physics Letters A*, 370(3):251–259, 2007.

[176] G.W. Wang and M.S. Hashemi. Lie symmetry analysis and soliton solutions of time-fractional $K(m,n)$ equation. *Pramana*, 88(1):7, 2017.

[177] A.-M. Wazwaz. The tanh method and the sine–cosine method for solving the KP-MEW equation. *International Journal of Computer Mathematics*, 82(2):235–246, 2005.

[178] A.-M. Wazwaz. The tanh and the sine–cosine methods for a reliable treatment of the modified equal width equation and its variants. *Communications in Nonlinear Science and Numerical Simulation*, 11(2):148–160, 2006.

[179] A.-M. Wazwaz. Solitons and singular solitons for the Gardner–KP equation. *Applied Mathematics and Computation*, 204(1):162–169, 2008.

[180] A.M. Wazwaz. Partial differential equations: Methods and applications. 2002. Balkema, The Netherlands.

[181] R. Weymann. Diffusion approximation for a photon gas interacting with a plasma via the Compton effect. *The Physics of Fluids*, 8(11):2112–2114, 1965.

[182] H. Xu. Analytical approximations for a population growth model with fractional order. *Communications in Nonlinear Science and Numerical Simulation*, 14(5):1978–1983, 2009.

[183] C.X. Xue, E. Pan, and S.Y. Zhang. Solitary waves in a magneto-electro-elastic circular rod. *Smart Materials and Structures*, 20(10):105010, 2011.

[184] N.N. Yanenko. Compatibility theory and integration methods for systems of nonlinear partial differential equations. *Proc. Fourth All-Union Math. Congr. in Leningrad*, 2:247–252, 1961.

[185] X.-J. Yang. *General Fractional Derivatives: Theory, Methods and Applications*. Chapman and Hall/CRC, 2019.

[186] J. Yin, L. Tian, and X. Fan. Classification of travelling waves in the Fornberg–Whitham equation. *Journal of Mathematical Analysis and Applications*, 368(1):133–143, 2010.

[187] M. Younis and S. Ali. Bright, dark, and singular solitons in magneto-electro-elastic circular rod. *Waves in Random and Complex Media*, 25(4):549–555, 2015.

[188] Z.J. Zawistowski. Symmetries of integro-differential equations. *Reports on Mathematical Physics*, 48(1-2):269–276, 2001.

[189] Xiang-Hua Zhai and Yi Zhang. Lie symmetry analysis on time scales and its application on mechanical systems. *Journal of Vibration and Control*, 25(3):581–592, 2019.

[190] J. Zhou and L. Tian. A type of bounded traveling wave solutions for the Fornberg–Whitham equation. *Journal of Mathematical Analysis and Applications*, 346(1):255–261, 2008.

[191] J. Zhou and L. Tian. Solitons, peakons and periodic cusp wave solutions for the Fornberg–Whitham equation. *Nonlinear Analysis: Real World Applications*, 11(1):356–363, 2010.

[192] Kai Zhou, Junquan Song, Shoufeng Shen, and Wen-Xiu Ma. A combined short pulse-mKdV equation and its exact solutions by two-dimensional invariant subspaces. *Reports on Mathematical Physics*, 83(3):339–347, 2019.

Index

Note: Page numbers followed by "*n*" indicate footnotes.